Alternative Energy in Agriculture

Volume II

Editor

D. Yogi Goswami

Professor
Department of Mechanical Engineering
North Carolina Agricultural and Technical State University
Greensboro, North Carolina

CRC Press, Inc.
Boca Raton, Florida

Library of Congress Cataloging-in-Publication Data

Alternative energy in agriculture.

Bibliography: p.
Includes index.
1. Renewable energy sources.   2. Agriculture and
energy.   I. Goswami, D. Yogi.
TJ808.A48   1986   630   86-4223
ISBN 0-8493-6347-0 (v. 1)
ISBN 0-8493-6348-9 (v. 2)

Direct all inquiries to CRC Press, Inc., 2000 Corporate Blvd., N.W., Boca Raton, Florida, 33431.

© 1986 by CRC Press, Inc.

International Standard Book Number 0-8493-6347-0 (Volume I)
International Standard Book Number 0-8493-6348-9 (Volume II)

Library of Congress Card Number 86-4223
Printed in the United States

PREFACE

Agriculture is highly dependent on energy, particularly on fossil fuels. Much of the fossil fuels used in agriculture can be saved through sound energy management and use of alternative energy sources. Research over the years has made available to us the methods of use of various alternative energy sources (such as solar energy, wind energy, methane, biogas, and alcohol), their design methodologies, and the design data needed. However, this information was scattered through thousands of publications. Therefore it was felt that there was a need to bring together all the available information for people interested in the study and practice of agriculture. This book presents comprehensive information on the use of various alternative energy sources in agricultural operations. The book contains the energy use patterns of various agricultural operations, availability of conventional fuels and utilization of alternative energy sources, such as solar energy, wind energy, methane generation, ethanol, and biomass gasification. The chapters on solar energy include fundamentals, use in water and space heating, drying, food processing, and photovoltaics. The information given about each alternative energy source includes fundamentals, potential for utilization, methods of utilization, and examples of operating systems and of designs. Data needed for the design of alternative energy systems is given in the text and in the appendixes.

The book will be useful for students, researchers, and teachers of agriculture and agricultural engineering, agricultural extension agents, and energy agents. It will also be useful to all other people involved in the design of alternative energy systems.

Recognizing that it is almost impossible for one author to write a book with such broad coverage, we selected authors for various chapters who are among the best in the world in their areas of expertise. They have done an excellent job of writing their chapters. I am extremely grateful to them. I am also grateful to Mrs. Dolores Ahrens for her help in coordinating this book and the editorial staff of CRC Press for their valuable assistance.

THE EDITOR

Yogi Goswami is a Professor of Mechanical Engineering at North Carolina Agricultural and Technical State University. He received his B.S. in Mechanical Engineering in 1969 from the University of Delhi in India and his Ph.D. in Mechanical Engineering from Auburn University in Auburn, Alabama in 1975. He worked as a Research Associate at Auburn University and as Assistant Professor at Tuskegee Institute before joining the faculty at North Carolina A & T State University in 1977 as Assistant Professor. He was appointed as Professor of Mechanical Engineering in 1985.

Dr. Goswami has been engaged in teaching, research, design, and consulting in the areas of thermal engineering, solar energy, and energy conservation. He has published numerous research papers in the energy area in national and international journals. In addition to this book, he has written a solar radiation handbook and edited an American Society of Mechanical Engineers (ASME) conference proceedings, *Solar Engineering—1984,* and a book entitled *Progress in Solar Engineering.*

Dr. Goswami has coordinated and chaired a number of national and international conferences on solar energy research and applications. He has served on a number of technical committees and boards of professional societies and has served on the Executive Committee of the ASME Solar Energy Division since 1984. He is a registered professional engineer.

CONTRIBUTORS

Volume I

Fred H. Buelow
Professor
Department of Agricultural Engineering
University of Wisconsin-Madison
Madison, Wisconsin

Barney K. Huang
Professor
Department of Biological and Agricultural
  Engineering
North Carolina State University
Raleigh, North Carolina

David E. Klett
Professor
Department of Mechanical Engineering
North Carolina Agricultural and Technical
  State University
Greensboro, North Carolina

Daryl B. Lund
Professor and Chairman
Department of Food Science
University of Wisconsin-Madison
Madison, Wisconsin

A. Shahbazi
Assistant Professor
Department of Plant Science and
  Technology
North Carolina Agricultural and Technical
  State University
Greensboro, North Carolina

Rakesh K. Singh
Assistant Professor
Department of Food Science
Purdue University
West Layfayette, Indiana

Volume II

R. Nolan Clark
Agricultural Engineer
U.S. Department of Agriculture
Agricultural Research Service
Wind Powered Irrigation Research Unit
Bushland, Texas

James Durand
Visiting Professor
Department of Mechanical and Aerospace
  Engineering
Columbia, Missouri

James R. Fischer
Associate Director
Agricultural Experimental Station
Michigan State University
East Lansing, Michigan

Eugene L. Iannotti
Associate Professor
Department of Agricultural Engineering
University of Missouri
Columbia, Missouri

Paul Longrigg
Solar Energy Institute
Golden, Colorado

J. Messick
Vice President
Belcan, Incorporated
Cincinnati, Ohio

Anil K. Rajvanshi
Director
Nimbkar Agricultural Research Institute
Phaltan, Maharashtra
India

Kyle Robinson
Independence Securities
Greensboro, North Carolina

TABLE OF CONTENTS

Volume I

TABLE OF CONTENTS

Volume II

Chapter 1

INTRODUCTION

D. Y. Goswami

TABLE OF CONTENTS

I. ENERGY AND AGRICULTURE

Agriculture is highly dependent on energy, particularly on fossil fuels. This is especially true of developed countries like the U.S., Canada, and Australia,[1,2] which provide most of the agricultural production of the world. Even in developing countries, where human and animal power are the main sources of energy on the farm, increases in the use of fossil fuels are taking place because of farm mechanization, which is essential to increase the food production needed to feed and clothe the ever-increasing population. Therefore, any decrease in the availability of or a substantial increase in the price of these fuels can have a devastating effect on the production of food and fiber.

The food system uses about 10 to 20% of the total energy used in the developed countries.[2] Of the energy used in the food system, 18 to 30% is used in agricultural production. In the U.S., agricultural production consumes approximately 3% of the total energy used nation-wide. However, this energy and its availability at critical times is of vital concern.

Various fuels used in agriculture include gasoline, diesel, fuel oil, LP gas, natural gas, coal, and electricity (generated from fossil and nuclear fuels). The table in Chapter 2, Section III.A shows the amount of energy used by type of fuel for U.S. agricultural production in 1974 and 1978.[3] As can be seen from this table, over 97% of on-farm energy use comes from fossil fuels. On the average, over 30% of the fossil fuels used in the developed countries are imported. Political uncertainties and conflicts among some of the oil exporting countries, as seen from events in the past, have increased the need for finding alternatives to the fuels that must be imported. The need to find alternatives to fossil fuels is there even for countries with their own reserves, because the fossil fuels will eventually be depleted.

II. ENERGY CONSERVATION AND ALTERNATIVE ENERGY SOURCES

Since the early 1970s, research on energy conservation and the use of renewable sources of energy has increased tremendously world-wide. Studies have shown that much of the fossil fuels used in agriculture can be saved through sound energy management practices and the use of alternate energy sources. Research over the years has made available to us the methods of energy conservation and use of renewable energy sources, their design methodologies, and design data needed. However, this information is scattered in thousands of publications (too numerous to refer to here). It was felt that there was a need to bring all the information available together for people interested in the study and practice of agriculture. However, it must be recognized that it is almost impossible to do justice to all the energy management techniques and the alternative energy techniques if one attempts to bring all of them together. Recognizing that some excellent sources are available for the various energy management techniques in agriculture (e.g., References 4 and 5), these two volumes were limited in scope to include only the important alternative energy techniques. The various alternative energy technologies included in these two volumes are

- Solar energy (thermal and photovoltaics)
- Wind energy
- Methane
- Alcohol
- Biomass gasification

The economic feasibility of any of these technologies depends on many factors, such as availability and costs of equipment needed for these technologies, type of fuels they can replace, availability at critical times, availability and costs of conventional fuels, etc. There-fore, while one technique may be economically feasible at one location, it may not be viable at another; if a technique is not economically feasible at a location now, it may become feasible a few years from now. The emphasis in these volumes is on the state of knowledge in the fundamentals, the design, and the applications of various alternative energy techniques.

REFERENCES

1. **Stout, B. A.,** Energy for world agriculture, *Energy Management and Agriculture,* Robinson, D. W. and Mollan, R. C., Eds., The Royal Dublin Society, Ballsbridge, Dublin, Ireland, 1982.
2. **Anon.,** The energy problem and the agro-food sector, Organization for Economic Co-Operation and Development (OECD), Paris, 1982.
3. **Torgerson, D. and Cooper, H.,** Energy and U.S. Agriculture 1974 and 1978, Statistical Bulletin No. 632, National Economics Division and Data Services Center, U.S. Department of Agriculture, Washington, D.C., 1980.
4. **Robinson, D. W. and Mollan, R. C., Eds.,** Energy management and agriculture, in *Proceedings of the First International Summer School in Agriculture,* Royal Dublin Society, Dublin, Ireland, 1982.
5. **Stout, B. A.,** *Energy Use and Management in Agriculture,* Breton, North Scituate, Mass., 1984.
6. **Ritchie, J. D.,** *Sourcebook for Farm Energy Alternatives,* McGraw-Hill, New York, 1983.
7. **Goswami, D. Y.,** Energy Conservation and Use of Alternate Energy Sources in Agriculture in North Carolina, Final Report, USDA Contract No. NC.X-020-5-79-440-4, North Carolina A & T State University, Greensboro, N.C., February, 1981.
8. **Anon.,** Energy in Agriculture — Research Needs in the Southern Region, Report of the Joint Task Force of Southern Region A.E.S. and the A.R.S., USDA, available from the Agricultural Engineering Department, University of Kentucky, Lexington, Ky., 1976, 40506.
9. **Anon.,** Solar Energy Applications in Agriculture Potential Research Needs and Adoption Strategies, Report NSF/RA-760021, Agricultural Experiment Station, University of Maryland, College Park, Md., January 1976.
10. **Brewer, R. N., Flood, C. A., Taylor, E. S., Koon, J. L., and White, M.,** *Solar Applications in Agriculture,* Franklin Institute, Philadelphia, Pa., 1981.

Chapter 2

PHOTOVOLTAICS USE IN AGRICULTURE*

P. Longrigg

TABLE OF CONTENTS

* The opinions expressed in this chapter are solely those of the author and do not in any way reflect D.O.E./ S.E.R.I. policy.

I. INTRODUCTION

The so-called energy crisis of the 1970s and its lingering financial consequences have made most people positively aware of the vulnerability of our energy-based economy. Nowhere has this struck home more than in the agricultural sector of the economy. Here profit margins are minimal and the rising costs of gas and liquid fuels can have a devastating effect. The number of farm bankruptcies recently in evidence no doubt have links, in many cases, to rapidly escalating energy costs, among other things.

In the last few years we have had a welcome respite to these ever-rising costs, due primarily to energy conservation. Nevertheless, the potential for supply disruption is still a distinct possibility, given the instability in places like the Middle East, Asia, and Central America. Under International Energy Agency rules, any overall shortfall in world production of oil must be shared on an equal basis by all consuming countries. Should this occur it will inevitably mean further price rises both for primary fuel and petroleum-derived products, like fertilizer. The already hard-pressed farmer will face extreme difficulty meeting such increased costs of operation. More bankruptcies could ensue as a result.

With ever-depleting reserves of fossils fuels, a case can be made for the use of renewable forms of energy in agriculture, despite, in some cases, marginal economics at this time. Nevertheless, in many instances some aspects of solar renewables are ruled out by high front-end costs. However, there are a few instances where the use of solar photovoltaics can be justified economically for agricultural use. This is particularly applicable in remote areas that do not have wired-in electric power, and where diesel-derived power is ruled out because of site accessibility and maintenance costs. Such remote applications might include groundwater desalination and pumping for crop irrigation, lighting, and boundary wire electrification. A further justification for the increased use of solar renewable energy is the effects on long-term climate change caused by the combustion of carbon-based fuels. Climate and the changes thereof are of vital importance to the agricultural community.

The National Science Foundation has called for increased efficiency in the world-wide use of fossil fuels together with an increasing adoption of plausible alternatives to fossil fuels as a way of mitigating the predicted global warming that will occur from the "greenhouse effect" of carbon dioxide gas released into the atmosphere by the continued combustion of hydrocarbons. Furthermore, our centralized electrical supply and distribution is highly vulnerable to both terrorist attack and instant destruction in a nuclear exchange. Decentralization of our electrical energy supply will decrease this vulnerability.

Agricultural applications similar to those noted above may be a good place to commence this deployment.

II. GENERAL ASPECTS OF ELECTRICITY USE IN AGRICULTURE

Extent of electric service — By June 1973, about 2,798,000 farms, or approximately 98.5% of the total number of farms reported by the U.S. Department of Agriculture (USDA) for that year, were receiving central station electric service. As of April 1973, the U.S. Census Bureau listed the farm population at 9,472,000, a reduction of 3,895,000 farm residents since 1963. Conversely, the average size of the farms has increased from about 322 acres in 1964 to 385 acres in 1974. Farm production has increased by an estimated 22% in 1973 compared with 1963, despite the 30% decrease in farm population. Farm output per man-hour increased by two thirds in the decade preceding 1973. Automation, mechanization, electrification, chemical fertilizers, better breeding, and management practices are credited with this change, but it has also involved greater energy use.

Energy use — The average per capita annual use of electricity on U.S. farms in 1973 was 12,588 kWh. This represents an increase of 46% in 10 years; it is expected to continue rising at a moderately accelerated rate.

Water heating — For water temperature usually requires (a rise of about 40°C), 1 kWh will heat 15ℓ of water (to this must be added all heat losses by radiation from tank or pipe). Storage tank heaters with tanks of 230 to 450ℓ capacity are sometimes used to obtain an off-peak rate when such is available. Under the more common policy of one rural rate for all energy consumed, when that rate is influenced by the customer's demand, a peak-limiting switch may be used to prevent the simultaneous operation of the water heater and the range. To be effective, instantaneous heaters usually have a demand of 3000 to 5000 W. Many tests under actual farm conditions in six northern states show an average monthly use of 289 kWh for farm household water heating.

Refrigeration — Freezing and refrigerated storage methods for processing and storing farm-produced food supplies have largely replaced such methods as pickling, brining, drying, and canning. Food freezers, home refrigerator storage rooms for cooling and preserving farm products, and the use of community frozen storage plants are popular in farm communities. The larger sizes of home refrigerators, with 0.3 to 0.35m³ of storage space and an average energy consumption of 40 kWh/month, are popular with farm families. Farmers prefer food freezers of the large type, with 0.3 to 0.6m³ of storage capacity. Average energy consumption varies from 3.56 to 6.27 kWh/ft³/month.

Irrigation pumping — More electrical energy is used for irrigation pumping than for any other field operation. Proper design of an irrigating system will depend upon the following factors: (1) the acreage and kind of crop to be irrigated; (2) the amount of water that must be supplied; (3) the amount of underground water available; (4) the depth at which water is found. Except where the water requirement is small and the depth to reach water great, plunger pumps are rarely used. The more common type is the centrifugal turbine pump, but where the lift is not more than 4.5m, the horizontal centrifugal pump is also used. The bowl of the turbine pump should be set below any expected drawdown in the well, and this will depend upon the porosity of the surrounding strata as well as upon the rate of pumping.

Vertical turbine pumps require vertical pumps with either solid or hollow shafts and thrust bearings capable of carrying the pump load. Horizontal pumps should be connected to their motors through flexible couplings to avoid the use of belts. With average allowance for evaporation, irrigation for an acre 0.3m deep requires $1.3 \times 10^6 \ell$. The soil can be made wet to a depth of 1.2m by using 100 to 150mm of water.[1] From 250 to 450mm is required to produce ordinary crops. With an overall efficiency of 50% for pump and motor, each acre-foot of water will require about 2 kWh of electricity per foot of water lift.

Methods of irrigation — Include overhead pipes, stationary spray plant, and portable sprinkler system. In the overhead type the discharge pipes are supported on ports and are located about 15m apart in lengths up to 180m. The pipes are usually supported on rollers so that they can be oscillated by a type of water motor, and nozzles are spaced 0.6m or more apart; 4.5ℓ/sec of water per acre at 14kg pressure is satisfactory. Stationary spray plants can reduce spraying time in orchards by 50% or more compared with portable units. A central pumping station, mixing tanks, and symmetrically located discharge pipes complete the layout. The pumps are usually three- or four-cylinder, single-action, with capacities of 1.5 to 4.5ℓ/sec at pressures up to 270kg or more, requiring motors of 5 to 30 hp. Outlets are located at regular intervals for attaching the spray hose. Spray nozzles discharge up to 0.6ℓ/sec depending on pressure and orifice size. Power required is usually under 10 kWh/acre of application. Portable systems utilize lightweight, quick-coupled pipes, with sprinklers attached. Laid on the ground, they require considerable labor to move, but the initial investment is less than with other types. Sprinklers operate at pressures of 140 to 350 KPa and cover circles 12 to 30m in diameter, delivering 0.2 to 2ℓ/sec. A motor as small as 2 hp will apply 25mm of water to 1.2 ha of land per week, although larger outfits are commonly used. These are standard practices of crop irrigation, using either surface water or ground-water, single- and three-phase wired electrical supplies. Very often, in remote areas such electrical supplies are not available or are extremely costly to lay in. It is in such a situation that a solar renewable electric supply can play a role. Furthermore, in many locations groundwater contains too high a level of dissolved salts for irrigation purposes, and desalination using electricity derived from solar photovoltaic systems can be used effectively.[2] This chapter will deal with these aspects of solar photovoltaic energy for water distribution and desalination for agricultural irrigation.

A. Energy from the Sun

The sun is a vast nuclear power plant of the fusion variety which generates power in the form of radiant energy at the rate of 3.8×10^{23} kW. An extremely small fraction of this is intercepted by the earth, but even this small fraction amounts to the huge quantity of 1.8×10^{14} kW, intercepted by the sun-side of the earth every 24 hr. The sun uses hydrogen for its fuel, which permeates all of space, and the nuclear reaction transmutes this hydrogen into helium, and in so doing releases radiant energy at X-ray, UV visible, and infrared wavelengths. It has been estimated that the sun has another 50 million years to go before it burns out and becomes a "white giant". During this time it will expand and eventually engulf the earth and all the planets in the solar system.

On the average, only about 60% of the Sun's power incident at the outer edge of the earth's atmosphere reaches the surface, or about 1.1×10^{14} kW. It must be remembered that this power is distributed over the entire surface of the sunlit half of the earth. What these numbers translate to is that for each square meter of surface (about 10 ft^2) 1 kW is incident. Thus, it can be seen that this energy is very diffuse in nature and with present state-of-the-art technology, expensive to collect.

To compare these numbers with our energy needs, consider that the present electrical-generating capacity in the U.S. is in the neighborhood of 5×10^8 kW. This is equivalent to the average sunshine falling on only 1000 mi^2 of territory in cloudless desert areas. Even if we could use only 10% of the incident sunlight, a single square plot of land of only 100 mi would provide the same electrical power as the present U.S. generating capacity.

Clearly, there is more than sufficient energy directly from the sun to meet both present and future needs. The trick is to learn how to gather and use it efficiently, and within normally accepted economic bounds.

B. Solar Photovoltaic Cells

Direct conversion of solar energy can occur in two ways. The incident sunlight can be turned directly into heat by photothermal conversion, using a device which selectively absorbs the rays of the sun. One method using this technique is the popular roof-mounted thermal array usually used for domestic hot water heating.

The incident sunlight can also be converted into DC electricity by the photovoltaic effect, using a device called a solar cell which utilizes the photoelectronic effects of a semiconductor junction, usually silicon.

An ever-growing group of manufacturers world-wide now make solar cells in large numbers. The cells are fabricated into multicell modules and these are then formed into arrays. Arrays are sized to meet customer needs. At present costs, the markets for these systems are almost exclusively for remote power applications. Such remote power applications will include water pumping and desalination as installing wired power from the utility grid increases in price.

Basically, a solar cell consists of a large area semiconductor diode so constructed that light rays can penetrate into the region of the diode p-n junction, usually a few microns (millionths of a meter) into the bulk material. Silicon is used almost exclusively for the material; however, other materials can be used and are showing considerable promise because of their relatively high efficiency.

To make a solar cell certain impurities are introduced into the silicon crystal lattice which allow positive and negative current carriers to exist when generated by sunlight. For example, phosphorus atoms give up valence electrons to the silicon to form n-type silicon (excess negative electronic charges) and boron atoms take up electrons from the silicon. One hole-electron pair is generated for each photon of light absorbed by the cell. Incorporation of impurities into the base material is called ''doping'' and is fundamental to the operation of the solar photovoltaic cell.

If photon absorption and hole-electron pair generation occurs near to the p-n junction formed by the doping, the electric field present at the junction will act to separate the holes from the electrons, causing holes to build up in the p-type material and the electrons in the n-type material. Connecting wires to the n- and p-type materials through an electrical load will result in current flow, as illustrated in the electrical circuit in Figure 1.

A simple analogy can be used to help explain how this happens. A solar cell can be likened to two tables with marbles on them, as shown in Figure 2. The two tables are connected by a small elevation or ridge over which the marbles are able to move if they have enough speed or energy imparted to them. Table A on the left is vibrated gently (in the horizontal plane) so that the marbles are jostled about, and occasionally a marble will roll over the ridge onto Table B. On the other hand, Table B is vibrated forcefully so that frequently the marbles will have enough energy to get over the ridge in the other direction to Table A. Eventually, as marbles accumulate on Table A and the number remaining on Table B diminishes, the number of marbles crossing the barrier will be the same in each direction, and condition of equilibrium is thereby established. However, if the two tables are connected together by an external channel, which we will assume for the sake of discussion to be flexible, the press of marbles on A will cause some of them to be pushed through this channel back to Table B.

In a solar cell, the p-n junction is like the ridge between the two tables. The free electrons

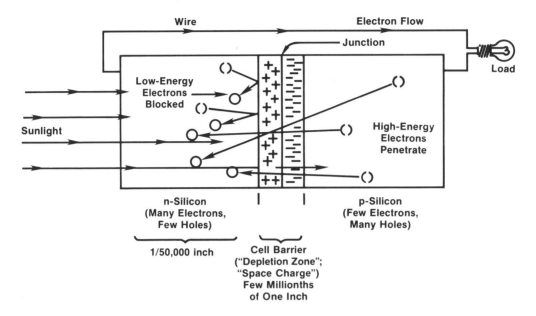

FIGURE 1.   PV cell schematic.

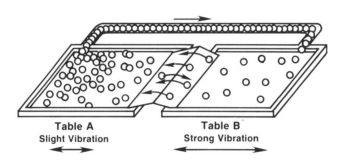

FIGURE 2.   PV cell analogy.

act like the marbles. The p-silicon lattice, corresponding to Table B, has an excess of holes due to the boron dopant and these holes tend to absorb electrons. As a result, there are not many free electrons in the p-silicon, but those that are free move at high speeds or energy levels. Figure 3 is an energy diagram of a p-n junction, which shows in pictorial form the energy required to move from the bound valence band of the silicon atom into the conduction band.

On Table A, corresponding to the n-type silicon, there are not enough holes and there are too many electrons, so that the average electron has less energy or is moving more slowly than those in the p-type silicon. Electrons in both layers move at random, some moving by chance into the electrostatic barrier caused by the p-n junction. The high-energy ones from the p-silicon penetrate the barrier into the n-silicon, while low-energy electrons in the n-layer are prevented from returning. Thus, voltage or electron pressure is created between the two layers. If a wire is connected externally with a load in series from the n-layer to the p-layer or vice versa, the excess electrons will readily flow through it.

As electrons migrate preferently into the n-layer, the holes in the lattice, of course, can be visualized as progressively "moving" in the opposite direction. Electrons coming back through the external wire eventually end up being absorbed in holes at the base of the p-

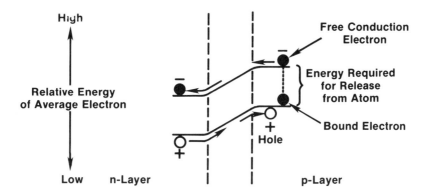

FIGURE 3.   Energy diagram.

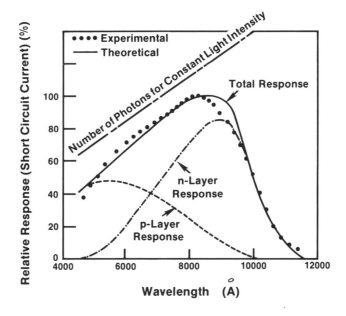

FIGURE 4.   Carrier creation within a solar cell.

layer, which in effect were originally produced when the electrons broke free. Electrons move so rapidly that if the source of light is shut off, the free electrons are almost instantly reabsorbed in holes and the process comes very quickly to a halt. The incoming light photons keep the process going by continually creating new electron-hole pairs. Indeed, it does not matter which side of the junction the new pairs are created on. It is important only that they be created in the vicinity of the junction. If a photon is absorbed in the n-layer, the electron that is freed simply adds to the number already present on that side, which is desirable; the hole migrates through the junction as incoming electrons successively fill it. If a photon is absorbed in the p-layer, the process is essentially as described above.

A representation of current carrier generation in a photovoltaic cell is shown in Figure 4; as can be seen, the current generation is very wavelength dependent. The n-layer is usually made of such a thickness that the junction will be at optimum depth for absorbing photons. For silicon, this is about 0.5 μm ($^1/_{50,000}$ in.).

The entire thickness of the cell need be no greater than 250 to 300 μm (0.25 mm) or about $^1/_{100}$ in. This is only three times the thickness of a human hair. It is theoretically

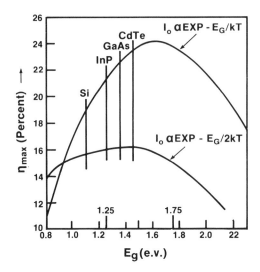

FIGURE 5.   Theoretical efficiency vs. semiconductor band gap.

possible to make solar cells that are extremely thin, and thus much cheaper, especially in mass production. Differing materials have different sunlight-to-electricity conversion efficiency, and Figure 5 shows some theoretical efficiencies for a number of materials, including silicon. The parameter E_g is the semiconductor band gap, which is the voltage difference of bound electrons and free electrons in the conduction band shown in Figure 3.

A scientist will use the diagram in Figure 3 to describe what happens in a solar cell. Here the cell is imagined to be in the position of the two tables of marbles described previously and shown in Figure 2, with the n-layer silicon on the left. Like the ridge between the tables, the junction or barrier is in the middle. The vertical dimension in the diagram is used to portray the relative energy of both the free electrons and the bound electrons in each region of the cell. The free or conduction electrons, represented by the upper line, move from the region of higher average energy in the p-silicon on the right, through the barrier, to the region of lower average energy in the n-silicon on the left. Simultaneously, the holes or vacancies migrate in the opposite direction.

Holes or electron vacancies in atoms may be considered in similar fashion to automobiles that move from the city to the suburbs in the evening. Expressed another way which is analogous to the atomic vacancy, an equal number of empty parking spaces "flows" into the city as the cars exit! If each driver had several errands to perform and therefore made a number of stops on the way home, occupying a number of parking spaces, then the picture of empty spaces flowing toward the city as cars flowed homeward takes on more realism.

Solar photovoltaic cells are unique energy producers in that no materials are consumed or given off. For this reason they can be completely sealed off from the atmosphere in optically translucent encapsulation materials. The fuel for these remarkable devices are the photons of light impinging on them, which replenish the supply of free electrons by energizing electrons from their trapped positions in valency orbits about their atomic nuclei, into a free state in the conduction band. Photons, in effect, provide a pumping action at the atomic level, and the energy they supply is converted into the energy of electrons in motion, and thus into electric current that can provide power for a light, drive a motor, or produce heat, as required.

It is important to differentiate photocells from batteries, for they are very different devices. The cell material in batteries must undergo change in chemical composition, during which the atoms, in the form of charged ions, move about. For example, lead may dissolve in acid as electricity is withdrawn and be plated out again as the cell is recharged. Because

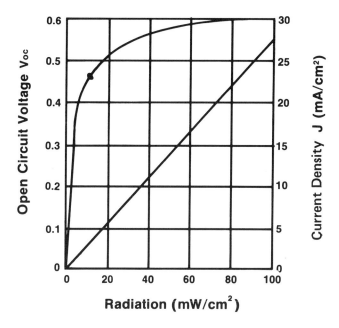

FIGURE 6. V_{max} and I as a function of light intensity.

the material itself changes in its chemical nature, it never is replaced exactly as it was, and even the best batteries wear out in time. Batteries work at the molecular level and are chemical devices. Solar cells work at the atomic level and theoretically, at least, never wear out in the classic chemical or mechanical sense. Nevertheless, they do degrade with time, but lifetimes are of the order of 20 to 30 years, and are primarily determined by encapsulant materials.

The amount of current or amperage produced by a solar cell is proportional to the amount of light falling upon it, i.e., the number of photons falling on it. For this reason, current will increase with the area of the cell as well as with the light intensity (see Figure 6). Furthermore, each cell can supply up to 0.5 V. At lower voltages the current supplied is nearly independent of voltage, but varies with light intensity.

Sunlight intensity has two components: the direct rays of the sun and the scattered or diffuse component (scattered by dust and gas molecules in the atmosphere). Solar cell arrays of the so called flat plate variety will respond to both components, called the global component, and therefore means that they will still work on cloudy days, albeit at somewhat reduced power.

With popular silicon solar cells about 160 mA of current can be obtained for each square inch of surface exposed to standard sunlight levels of 1000 W/m. The maximum power delivered to an external load is typically 11 to 12% of the total solar energy incident on the cell. Efficiencies of up to 15% have been observed on some laboratory models, which is not too far off the theoretical maximum efficiency of 18 to 20% as shown in Figure 5.

The above-mentioned levels of currents and voltage appear at first sight to be miniscule compared, say, to the line voltage and current consumption of typical loads; with solar cells it is possible to connect them in series to obtain higher voltages, and also in parallel to obtain higher load current. These concepts have been carried out successfully to the megawatt level on a number of electrical utility applications of photovoltaics. Voltage and current requirements are connected when the individual cells are packaged into modules. The modules then become the building blocks for arrays designed to meet a customer's specific needs.

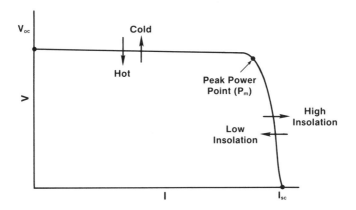

FIGURE 7.   Typical solar photovoltaic array characteristic.

An important characteristic of solar photovoltaic cells is their temperature sensitivity; they become less efficient converters of sunlight as their temperature rises. This must be taken into account when a PV system is designed. Figure 7 shows the effects of temperature in qualitative form. Solar cells perform best on cold clear days and least well on hot cloudy days.

The quantitative performance attributes of PV cells are adequately described by an I-V curve, or a curve that relates incident radiation to cell voltage and current with temperature as a parameter as shown in Figure 7. Ideally the cell should be operated at its maximum power point under all incident light conditions and electrical loading. This point is denoted as P_m in Figure 7. To achieve this condition, it is necessary to employ relatively complex circuitry in the power conditioning system that tracks maximum power point, as it changes with weather and diurnal variations.

In order for the solar cell to work properly, the surface must be accessible to incident light. Therefore, the exposed surface cannot be covered by an opaque or other similar material. However, a metal contact fixed to the surface is needed to pick up the photocurrent, so that at least part of the surface is covered with this opaque, electrically conductive material. Thus, the trick is to cover as little as possible of the surface while providing sufficient current-carrying capacity and widespread distribution of pickup points over the surface to avoid any signficant power dissipation in the contacts. For this reason, the current contacts are put down in the form of a grid as shown in Figure 8. Voltage is taken off the front and back contacts. Work in some research laboratories is at present centered on improving efficiency and cost of solar cells, and also the use of optically translucent, electrically conductive material, like indium (or cadmium) tin oxide (ITO) for current gathering.

The ''goodness'' parameter of a solar cells is termed its ''fill factor'' and a cell has a fill factor equal to unity when the ordinate to the maximum power point is equal to the abscissa to the same point, or in other words, the I-V curve is a square of maximum area under the I-V curve. This is, of course, the ideal case and never exists in practice.

III. CALCULATION OF ENERGY OUTPUT FROM A PHOTOVOLTAIC SYSTEM

In order to calculate the annual energy output given by a solar photovoltaic installation, it is necessary to have an elementary understanding of several performance parameters.

A. Performance Characteristics

The primary parameters that quantitatively describe the performance of a photovoltaic

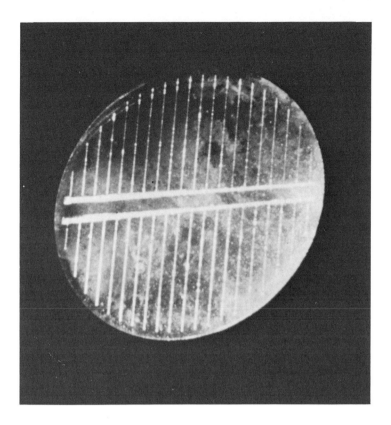

FIGURE 8.   Current contacts on photovoltaic cell.

device are the current and voltage (I-V) curves as shown in Figure 9. Current is proportional to the solar irradiance falling on a photoelectric surface, and the spectral distribution response. Voltage is sensitive to temperature but the three quantities — voltage, current, and power — have temperature coefficients. The P_{max} point on the I-V curve for a given temperature is that point which maximizes the area ABC.

As the maximum power point is liable to change as a function of solar irradiance and atmospheric conditions, arrangements may be needed to have the load track the locus of the maximum power point as shown in Figure 9. This is termed maximum power point tracking (MPT), and circuitry to do this is often incorporated in the power conditioning equipment. Use of MPT techniques will give between 3 to 5% energy recovery from the PV system, and because of the added complexity and its effect upon reliability, this recovery is considered by some to be marginal compared to variable power operation.

When calculating yearly average energy output from a photovoltaic installation, a set of reference conditions is needed to work with. These are

1.   Solar irradiance level (I_{total})
2.   Normal operating cell temperature (NOCT)
3.   Wind speed (WS)
4.   Air mass (AM)

For peak conditions the I_{total} is 1000 W/m² at an AM of 1.5³ and a NOCT of 25°C. Under

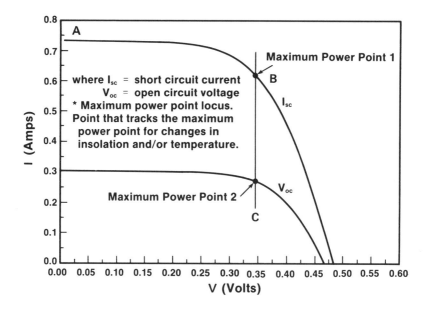

FIGURE 9.   Photovoltaic current-voltage curve with maximum power point.

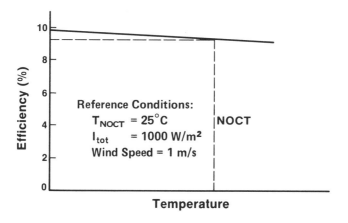

FIGURE 10.   PV efficiency vs. temperature.

nominal operating conditions, the I_{total} is 1000 W/m^2 with an AM of 1.5 and a NOCT of 25°C. The NOCT can be found by using the following regression:

$$T°C_{(cells)} = T_{ambient} + 0.03 \cdot I_{total} \qquad (1)$$

NOCT is defined as the PV cell junction (see Figure 10) temperature operating in thermal equilibrium with its environment.

A reference WS of 1 m/sec is assumed, and the efficiency of the PV device is needed under reference conditions. Efficiency under reference conditions can be calculated thus:

$$n_{eff} = \frac{P_{max}}{I_{tot} \cdot A} \qquad (2)$$

where A is the module/array area which can be obtained from manufacturer's data. Reference data curves of manufacturer's data usually have the format seen in Figure 10.

The total yearly integrated irradiance available to an inclined-plane, fixed-angle flat plate PV system, incident at the location of the PV installation (in kilowatt-hours per square meter), must also be available for the yearly energy calculation as given below:[4]

$$P_{system} = N \cdot P_{max} \cdot (e_1) - E_{aux} \tag{3}$$

where N = number of modules within array to meet load requirements. (N = load wattage requirement divided by peak watts of module containing series and parallel cells, as noted); P_{max} = maximum peak power of module.

$$e_1 = e_{mismatch} \cdot e_{wiring} \cdot e_{power\ conditioner}$$

where $e_{mismatch}$ is the reduction in module efficiency due to cell mismatch error (this has been assigned a value of 0.96 when quality control procedures include cell and model I-V curve matchings; without such procedures a value of 0.90 should be assigned to this parameter); e_{wiring} is the reduction in efficiency due to voltage drop along cabling and wiring (this parameter will typically be 0.99 or better for good electrical wiring practice that follows N.E.C. Article 690); $e_{power\ conditioner}$ is the efficiency of the power conditioning unit (this parameter will vary dependent upon the level of power being handled; the manufacturer should be able to supply efficiency figures for his power conditioning equipment at 25, 50, and 75% of rated full load; the average of these figures should be used); E_{aux} is the energy requirement of auxiliaries. For a remote flat plat system this will typically be for controls having a constant energy drain on the system and is assigned a value of 0.99.

The efficiency factors in Equation 3 assume a structurally integral array, i.e., one piece, and not separately mounted modules that could shadow each other. If the latter is the case, allowance* should be made for it.[5]

Furthermore, in Equation 3 the product $N \cdot P_{max}$ may be replaced with the expression:

$$n_{ref} \cdot A_{total} \cdot I_{ref} \tag{4}$$

The PV systems total annual energy output is calculated thus:

$$E_{sys(a)} = n_{ref} \cdot A_{total} \cdot S_{ref} \cdot (e_1) - E_{aux} \tag{5}$$

where A = total area of array and S_{ref} = total integrated yearly solar irradiance available on site, other parameters having been defined previously.

IV. GENERAL PV ARRAY SIZING METHOD

The basic assumptions for sizing the solar cell array are as follows:

1. The typical solar cell used supplies 1.1 A per sun. One sun equals 1000 W/m.[2] Therefore, a typical solar cell supplies 0.0011 A m²/W.
2. The typical solar cell supplies $1/3$ V per cell.
3. The typical solar cell supplies $1/2$ peak W per cell.

The daily amp-hour (DAH) is calculated by multiplying the insolation on the array tilted surface by 0.0011 A m²/W.

---

* Inter-array shadowing. For flat plates with a length to height ratio greater than two, or for all modules contained in a single plane, this factor equals 1. For multiple modules of length to height ratio = 1.5 and 2.0, the factor equals 0.95 and 0.99, respectively.

$$DAH = \text{insolation in } \frac{W \ hr}{m \ day} \times 0.0011 \ \frac{A \ m^2}{W \ cell} = \frac{A \ hr}{day \ cell} \qquad (6)$$

Solar cells in series add their individual voltages, and solar cells in parallel add their individual currents. The following equation is used to find the number of solar cells needed in parallel to supply the required current.

$$ARRAY = \frac{\text{load (W hr)} \times 1.3}{\text{(day) system voltage (V)} \times DAH \ \frac{(A \ hr)}{day \ cell}} = \frac{\text{number of}}{\text{parallel cells}} \qquad (7)$$

where ARRAY = number of parallel cells needed to meet current requirements.

$$LOAD = \text{duty cycle 1 (hr)} \times W \ 1 \ (W) + \text{duty cycle 2}$$
$$\times W \ 2 + \text{duty cycle 3} \times W \ 3 = \text{total W hr/day}$$

1.3 is a 30% safety factor to take into account cell degradation with age, dust, and loss of contacts.

The number of cells needed in series is equal to the system voltage times three.

$$SERIES \ CELLS = \frac{\text{system voltage (V)}}{1/3 \ V/cell} = \text{number of series cells} \qquad (8)$$

The total number of solar cells needed is equal to the number of parallel cells times the number of series cells. The total installed peak wattage of the solar cell system equals the total number of cells times one half peak watt per cell. The total installed peak watts is required when buying solar cells and is how solar cells are priced. The total peak wattage of the solar cells system will be several times the load wattage because the solar cell system runs at its peak only at noon on clear days.

A. Energy Storage

The total energy received from the sun in 1 year at any particular location is surprisingly constant. The variation from year to year is less than 10% in most places. However, within 1 year, large variations can occur both seasonally and daily, much of which is unpredictable. To avoid power outages at night and during inclement and heavily clouded weather, some means of storing the energy generated during sunny weather can be employed. All photovoltaic generating systems in use today at remote locations without wired-in utility power employ rechargeable electric storage batteries for this purpose.

The solar generation array is designed and sized to provide all the power required by the load all year round. In this system the storage battery acts as a buffer between the weather-conscious solar array and the constant power input requirements of the load. They supply power to the load during periods of low sunlight or no sun at all and accept charge from the photovoltaic array during periods of high sunlight when generation is in excess of load demand.

Therefore, the state of charge may vary seasonally as well as daily, as shown in Figure 11. The average sunshine during the winter may be insufficient to supply the load, but this insufficiency will be made up by an excess of sunshine during the summer.

The capacity of the storage battery must be chosen to meet these demands. The storage battery also serves conveniently to fix the voltage at which the system will operate. These and other requirements put special emphasis on the battery performance specification. The standard automotive lead acid battery is *not* optimum for this type of service.

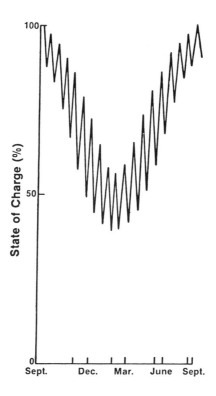

FIGURE 11.   Solar-power battery-state of charge profile.

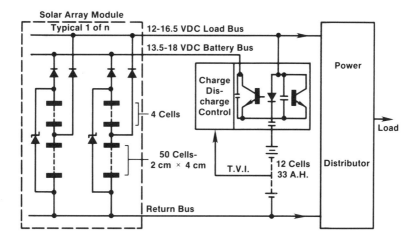

FIGURE 12.   Zener regulated solar array/battery system simplified schematic.

B. Solar Photovoltaic Energy Generator Systems

In addition to a properly sized solar cell array and storage battery, a blocking diode must also be incorporated. This is included to prevent battery current drain through the solar cells at night. In complete darkness the solar cell array is simply a string of diodes which will be forward biased by the storage battery. Thus, the blocking diode between the solar array and the battery, shown in Figure 12, allows current to flow to the battery from the array but prevents current flow from the battery to the array.

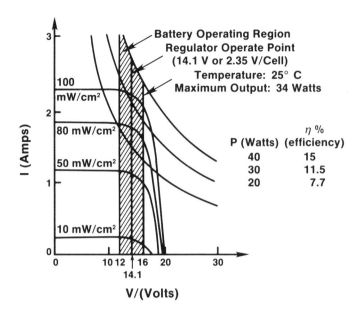

FIGURE 13.   Typical solar module characteristics curves with battery characteristics.

A voltage regulator or charge controller may also be included in a complete system, similar to that shown in Figure 12, to avoid overcharging the battery, which results in a loss of the electrolyte of the battery. This will occur when the battery is fully charged, but the solar array is still supplying more power than the load requires. Under these conditions, the voltage regulator will shunt the excess current through a dummy load which could be an electric heating element. A voltage regulator may not be essential for systems which are not expected to supply a large excess of energy over the year, and if the operator is willing to visit the site to replace any lost electrolyte. However, a voltage regulator is recommended for all systems which are expected to produce over 20% excess energy over the year.

C. Solar Cell Module Design

It is common practice to describe the size of a solar photovoltaic array in terms of its peak power output, which is the maximum power that can be delivered to a matched load when the incident sunlight has an intensity of 100 mW/cm^2 for a given air mass that the sunlight traverses to get to the array. This latter specification generally defines the time of day, which is nearly always solar noon, or meridian.

However, by itself, this is not a significant parameter in the design of a solar array-battery charging system. Both the current and the voltage characteristics of the solar array must be considered in conjunction with the battery voltage and charging characteristics to obtain a correct match between the two, as shown in Figure 13. This means that power must be supplied to the battery at a high enough voltage to fully charge the battery. For a typical 12-V lead-acid battery this voltage can range up to slightly over 14 V.

Current-voltage (I-V) curves for a typical solar cell module with 100 mW/cm^2 incident sunlight are shown in Figure 14. Both curves are taken from the same module, but the temperature of the cells in the module is different. The third curve in Figure 14 is the product of voltage and current which, of course, is power. The maximum power point on each curve is indicated by a point. This particular module contained 36 cells of approximately 100 mm diameter, all connected in series, and is designed for charging a 12-V lead-acid battery.

It should be noted that the solar cell module acts essentially as a constant current source

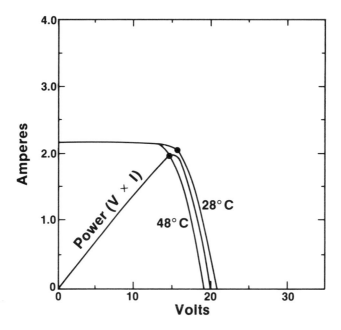

FIGURE 14.   I-V curves of a typical solar module.

nearly independent of voltage below the maximum power point, but the current falls off rapidly with increasing voltage above the maximum power point. The voltage at which this falloff occurs depends on the number of cells connected in series, and also upon their temperature.

By the choice of 36 cells, the module can supply a fully rated current of 2 A to the battery even when it is almost fully charged at 14 V charging voltage and at cell temperatures up to 60°C. With minimal voltage drops across the blocking diode and wiring cable (i.e., 0.25 to 0.35 V), this allows the nominal current rating of the module to be used in sizing systems without temperature derating for most applications.

Although an additional cell in series or a 37-cell module would cause the current drop-off to occur at higher voltage, and increase the power rating of the module and its cost, no real benefit would be derived for charging a standard 12-V battery.

As shown in Figure 7, the voltage fall-off occurs at lower voltage as the temperature of the solar cell increases. The amount in fall-off voltage is given approximately by 0.5% per degree Celcius for silicon solar cells. By this means, temperature derating of modules can be taken into account when sizing systems.

In most applications, weather conditions are such that cell temperatures will rarely exceed about 69°C, and most modules from reputable firms are designed for these normal applications. However, applications are sometimes encountered where climatic conditions can lead to cell temperatures reaching 65 to 75°C. For example, in low-altitude tropical locations, ambient temperatures may reach 40 to 50°C in the hot season. Under these high temperature conditions, cell temperatures will often exceed ambient by 30 to 35°C depending upon the type of module encapsulant used and site wind conditions encountered. Under these conditions the derating of modules should be taken into account when systems are sized. Usually the result will not affect the system design because this derating occurs at the time of the year when the battery has been brought to full charge by excess output from long hours of sunlight. Therefore, there is no justification to design a special module with one or two extra cells in series for these special applications.

The constant charging current nature of a photovoltaic solar cell array properly matched to the storage battery system allows system sizing to be easily computed in terms of ampere-hours. This is very convenient because this quantity is very nearly conserved in a storage battery. Ampere-hours 'out' almost equals ampere-hours 'in', even though power out is less than power in, due to the less than 100% efficiency of the battery, which is manifest by amps out at lower voltage than they go in. In addition, the current supplied by the solar array is more or less directly proportional to the incident sunlight. This does not affect the constant current characteristic of the solar array with respect to voltage, because the ordinate or short circuit current of the array moves up and down as a function of the incident sunlight. This feature leads to a direct correspondence between integrated sunlight data and total ampere-hours supplied by the array.

The current level obtained from a module of series-connected cells, as shown in Figure 12, is the same as that obtainable from a single cell in the module; this, in turn, is fixed by the exposed surface area of the cell. To obtain lower current levels than those discussed here (i.e., 2 A), a module with smaller-diameter cells is required. Current limiting would not, of course, be economic. However, the number of series-connected cells to match a given storage battery is independent of solar cell size. Higher current levels can be obtained by either using larger cells, if they are available, or by connecting modules in parallel, which is often done. In the latter case, the total current obtained is the sum of the currents from the individual modules. This does not change the voltages at which the current drops off, which is fixed by the number of cells in series in each module. For example, two electrically identical modules connected in parallel would give identical curves to that shown in Figure 14 if the current scale were doubled.

Varying battery voltages are accommodated by connecting modules in series. If one module is properly sized in voltage for a 12-V storage battery then two such modules in series connection will be correctly sized for 24-V battery operation. Thus, a specific application may require modules to be connected both in series and parallel: series connected to match specific battery voltage, and additional modules connected in parallel to provide the required current generating capacity. An example of these techniques is given elsewhere in this chapter.

Sizing methods for solar photovoltaic cells is generally predicated on the fact that module performance I-V is similar to those shown in Figure 14. However, other solar cells may give rise to different I-V curve shapes which can modify the arguments for the number of series-connected cells needed to charge a 12- or 24-V battery.

In Figure 15, curve A is a repeat of the solid line shown in Figure 14. If this module had been constructed of cells having relatively high current leakage, the I-V curve would look something like that shown in curve B of Figure 15. In this case, the nominal current rating of the module would have to be specified at a lower value, and conceivably it might be advantageous to put additional cells in series to minimize further derating at higher cell temperatures.

If the module were to be constructed having excessive series resistance, the I-V curve would be something like curve C in Figure 15. In this case, it is clear that additional series-connected cells would be advantageous to reduce very large derating which this module would exhibit at higher cell temperatures. Furthermore, the direct correspondence between integrated sun insolation data and total ampere-hours supplied by the module would no longer be independent of sunlight intensity, unless the number of series-connected cells were increased by about 25%. This would adversely affect the cost. *It is therefore vital to have a knowledge of the shape of the module I-V curve for proper module and system design.*

Normally, cell performance characteristics are matched fairly closely before being encapsulated into a module. However, performance degradation can occur over time due to the natural aging and discoloration of encapsulant materials. Consequently, yearly or biyearly checks should be made on overall system performance. The aim of much current research is to enhance lifetimes of encapsulating materials.

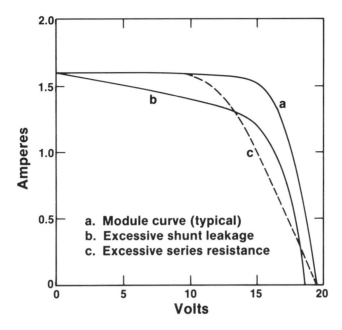

FIGURE 15.   Effects of shunt leakage and series resistance on I-V curves.

D. Storage Batteries for Solar Photovoltaic Systems

The major factors influencing the adoption of storage batteries for use with solar photovoltaics are the battery type required, the voltage, ampere-hour capacity, and the environment in which it will be operating.

The two most common types of batteries used in solar electric systems are lead-acid and nickel-cadmium types. It is important in the case of lead-acid batteries that low self-discharge types of batteries be used. For remote applications they should also be capable of periodic deep discharge. For lead-calcium or pure lead grid constructions, losses are generally 1% of rated capacity per month. Furthermore, lead-calcium types are preferred due to their higher mechanical strength. Lead-antimony grid types are not recommended since their initial self-discharge is 7 to 8% of rated capacity per month, and as the battery ages this figure rises to as much as 30 to 40% of capacity per month.

Self-discharge loss becomes a load to the photovoltaic array and will increase the size of the solar panel without providing any power benefits to the user. Battery capacity must be specified for the environmental conditions to which the battery will be subject. Batteries are normally rated at normal temperatures, i.e., between 0 and 30°C. When a battery cannot be isolated and/or insulated from abnormally cold temperatures (e.g., -20 to -40°C) it is necessary to derate its capacity. It cannot be allowed to discharge to the point where the electrolyte will freeze. This derating factor can be obtained from the known variation of freezing temperature with specific gravity of the electrolyte.

Another factor influencing the selection of a battery is its charging efficiency. Up until the battery reaches 90% of full capacity, the solar panel charges the battery essentially in a constant current mode. Therefore, at this juncture we need only be concerned with ampere-hour charging efficiency of the battery. This efficiency is typically 95% acceptance or greater at normal temperatures. The bulk of the charging inefficiency manifests itself in the form of a voltage drop. The battery requires a higher voltage to accept a given number of ampere-hours than it provides when the same ampere-hours are withdrawn. Also, as the battery cools down, the charging voltage tends to increase.

As the battery reaches and exceeds 90% of capacity, the array current decreases rapidly with increasing voltage. If the solar panel voltage (i.e., the number of cells in series in the module/array configuration) and current outputs have been carefully matched to the load, this will provide a self-regulating mechanism. However, when the solar array outputs exceed the load requirements by more than 20 to 30%, an external voltage regulator, similar to that shown in Figure 12, may be required to avoid excessive overcharging of the battery. Such a situation can arise when attempting to accommodate a variable load by adding more modules to the array already fielded.

Batteries used with solar photovoltaic systems often operate in conditions that favor stratification of the electrolyte in that they are in stationary service. As charging currents are typically low in this application, occasional gassing may be permitted to break up the stratification, but maintaining electrolyte level as much as possible, thus dictating a very low level of gassing.

The other main type of battery used in remote solar photovoltaic applications is the nickel-cadmium type. This battery has a low rate of self-discharge and is more tolerant of misuse than is the lead-acid type. For example, they will recover fully from freezing conditions, and discharge to zero. Excessive overcharging will not damage the battery, but water is lost and must be replaced. These advantages are offset by high unit cost, voltage inefficiency typically requiring a charging voltage 15 to 20% higher than the discharge voltage, as compared to 5% for lead-acid batteries. Furthermore, they have lower voltage per cell than the lead-acid, and suffer from discharge "memory". This latter aspect is a refusal to discharge to a level lower than that to which it had previously been repeatedly discharged.

V. CHARGE CHARACTERISTICS

The amount of water loss in a lead-acid cell is proportional to the ampere-hours of overcharge. Thus, in cells designed for minimum maintenance, the overcharge must be kept to a minimum. If it is too low, however, the acid stratification developed on discharge will not be completely removed and a decline in capacity will result. The optimum overcharge for low-maintenance industrial cells was investigated in detail and some of the results are reported below. Low-maintenance cells of 400 ampere-hours nominal capacity were used to investigate (1) the relationship between stratification, overcharge, and capacity; (2) the amount of gassing on charge; (3) the influence of cell voltage during overcharge on the charge current.[6]

As is typical in industrial cells of the motive power type, the cell element, comprising alternating positive and negative plates and separators, fits tightly into the cell container leaving very little space for free acid at the sides. Above and below the cell element there is about 2 cm of free sulfuric acid electrolyte.

VI. STRATIFICATION

Stratification is the term used to describe the vertical concentration gradient of sulfuric acid that develops during discharge. It is caused by the dilution of H_2So_4 at the positive electrodes. The dilute acid that is formed in the pores rises to the top of the cell and is displaced by denser acid in the space between the plates. This convective flow results in a density gradient over the height of the cell.

A. Gas Evolution on Charge

The round-trip coulombic efficiency of a lead-acid battery is typically about 85%. This efficiency is less than 100% due to the occurrence of oxygen evolution at the positive and hydrogen evolution at the negative before the discharge ampere-hours are replaced. Consequently, a degree of overcharge is needed to fully recharge a lead-acid cell. Overcharge is also needed to remove stratification as discussed above.[6]

VII. CHARGER DESIGN

The data on charge characteristics reported above were used in designing an improved charger for use with low-maintenance (and conventional) deep-cycle batteries. The charge profile used in the laboratory studies was similar to that obtained with the commonly used ferroresonant chargers. A major problem with using conventional ferroresonant chargers with low-maintenance batteries is that the overall charge time is fixed on the assumption that batteries will normally be deep-discharged. This means that batteries will receive an amount of overcharge in inverse proportion to the previous depth of discharge. To overcome this problem we need a solid-state control circuit to replace the electromechanical timer. This circuit can be retrofitted to existing chargers. The circuit operates by sensing a preset trip voltage (approximately 2.42 V per cell). Upon tripping, a timer is started and charging continues for a fixed period. With this arrangement, the initial high-rate charge period varies in proportion to the prior depth of discharge and the ampere-hours of overcharge is independent of the depth of discharge.

The maintenance costs of conventional industrial batteries over their operating life are similar to the initial cost. To reduce these costs and improve the reliability of such batteries, a line of low-maintenance lead-acid batteries suitable for both deep- and shallow-cycle applications has been developed. When deep-cycled on a daily basis, these batteries need water addition every 100 to 125 cycles compared to every 5 to 10 cycles in the case of conventional industrial batteries. The low-maintenance characteristic is achieved by the use of a positive grid alloy that contains only 1.5% Sb coupled with an Sb-free negative grid alloy. This hybrid grid alloy combination has all of the advantages of an Sb-free combination (low gassing on overcharge, stable gassing characteristics over life, excellent capacity retention on open-circuit stand) and none of the disadvantages (low active material utilization, relatively poor deep-cycle performance). The features of this new low-maintenance battery are discussed with emphasis on grid alloy and charge characteristics. In tests with 400 ampere-hour cells, it was found that the optimum charge regime for minimum water loss and maximum capacity retention is 5% overcharge at a maximum cell voltage of 2.55 V with a periodic 10% equalization charge.

All secondary batteries must, of course, be recharged, but the historically low cost of electrical energy has provided little incentive for the improvement of charger efficiency and for the better matching of chargers with batteries. This lack of concern with chargers has meant that lead-acid batteries tend to be excessively overcharged, especially after being subjected to shallow discharges. Excessive overcharging not only wastes electrical energy but also accelerates electrode deterioration through corrosion, active material softening, and shortens the period between water additions.

VIII. CHOICE OF GRID ALLOYS

The amount of gassing, and hence water loss, on charge is controlled by the composition of the grid alloys. Unalloyed lead is too soft and has insufficient creep resistance to be used as a grid material unless the plates are assembled horizontally rather than vertically. *Lead-acid batteries designed for deep-cycle use are normally manufactured with grids having antimony contents in the range of 4 to 6% by weight.* The presence of antimony is desirable for rapid casting and to ensure reliable cycling performance of the positive (PbO_2) electrode. However, antimony has a low overvoltage for H_2 evolution at the negative (Pb) electrode. Hence, conventional batteries tend to lose material from the positive and accumulates by electrodeposition at the negatives.

One solution to this antimony problem is to use antimony-free grids such as Pb-Ca-Sn. These are widely used in batteries subjected to shallow or infrequent deep discharges such as maintenance-free automotive batteries and most stationary or standby batteries. The disadvantage of such alloys is that, when used in the positive, the plates have lower specific capacities (ampere-hour per kilogram) and a less reliable and shorter cycle life than plates with antimonial grids. The reason for the beneficial effects of antimony on capacity and on cycle performance is still unclear but it appears to be related to the effect of antimony on the morphology of the active material (PbO_2) at the grid-active material interface.

Another solution is to use a positive low-Sb grid alloy with an Sb-free (Pb-Ca-Sn) alloy in the negative. This hybrid construction was initially developed for use in maintenance-free automotive batteries. Extensive laboratory and field testing showed that batteries fabricated with these grids had current-voltage, gassing, and water loss characteristics similar to batteries with Sb-free grids. Moreover, the batteries had cycling characteristics similar to batteries with antimonial grids. Thus, the antimonial grid is ensuring good cycle performance at the positive electrode while the antimony-free grid is minimizing the tendency for hydrogen to evolve at the negative electrode.

The low-antimony alloy referred to above is a ternary lead alloy containing 1.0 to 2.0% Sb and 1.2 to 2.2% Cd.[6] This alloy has excellent metallurgical and electrochemical characteristics and is currently being used in low-maintenance, deep-cycle batteries.

IX. ENERGY REQUIREMENTS FOR WATER PUMPING[7]

The commencement point for any analysis of water pumping energy requirements is to establish the relationship between water volume requirements and energy needed. The formula connecting these two parameters is

$$E = pgVh \qquad (9)$$

where E is the hydraulic energy in Joules (J), V is the required volume of water in cubic meters (m^3), h is the total head of water in meters (m), p is water density standard (1 million grams per cubic meter), and g is acceleration due to gravity (9.81 m/s^2). When V is given in cubic meters and h in meters the pumping energy requirement is

$$E = \frac{9.81 \ Vh}{1000}, \ MJ \qquad (10)$$

Thus, to lift 50 m^3 of water through a head of 20 m requires:

$$(9.81 \times 50 \times 20/1000) = 9.8 \ MJ \ or \ 2.7 \ kWh$$

The power (P) required to lift a given quantity of water depends on the length of time the pump is required. As power is defined as the rate of doing work or expending energy, the formula for hydraulic power is simply obtained from the formula for energy by replacing volume with flow rate (Q) in cubic meters per second thus:

$$P = pgQh \ W \qquad (11)$$

If the flow rate (Q) is in liters per second then the hydraulic power is

$$P = 9.81 \ Qh \qquad (12)$$

For example, the average hydraulic power required to lift 60 m^3 of water through 5 m of head in 8 hr, or an average flow rate of 2 ℓ/sec is

$$9.81 \times 2.08 \times 5 = 102 \ W$$

Using a typical pump efficiency of 70%, the mechanical power required is

$$102/0.7 = 145 \text{ W}$$

In water pumping operations it is energy (E) that is the more important parameter, since it is energy that has to be paid for in the form of diesel fuel, human and animal effort (acting over time), or solar pump size. The equivalent power requirement only determines how quickly the required quantity of water is delivered and the rate at which the energy is used.

Work is at present proceeding in standards-setting organizations to evolve a method for specifying energy rating of photovoltaic arrays at specific geographic locations under solar irradiance average conditions at a site.

The head (h) has a proportional effect on the energy and power requirements with the result that it is cheaper to pump water through lower heads. It consists of two parts: static head, or height through which the water must be lifted against gravity, and the dynamic head, which is a pressure increase caused by friction of water flow through pipework expressed as an equivalent head of water. Static head is easily determined by simple measurement. Dynamic head depends upon flow rate and pipe sizes and materials used. The higher the flow rate and the narrower and rougher the pipe inner surface, the greater the dynamic head, and therefore the pressure needed to force a given amount of water through the pipe work.

Materials used for the pipework are very important in terms of life cycle costing. Water temperature and dissolved oxygen content are important parameters to consider when black iron pipe is contemplated. Under high temperature and dissolved oxygen content conditions, low pH conditions often prevail with a consequent leaching of the iron by carbonic acid.

As a general axiom it can be shown that solar pumping systems for irrigation purposes are starting to be cost-effective compared to diesel-powered pumping in applications where the peak daily water requirements are less than about 150 m^4,* e.g., 30 m^3/day brought through a head of 5 m, and where the minimum monthly average solar irradiation is greater than about 15 MJ/m^2/day.

For rural water supply (human and animal stock) requirements, solar pumping systems are cost-effective where average daily water requirements do not exceed about 250 m^4, e.g., 25 m^4 through a head of 10 m and when the solar irradiation incident is greater than 10 MJ/m^2/day.

Many parts of the world meet these conditions and they are generally in areas where the water supply situation is often critical.

Figures 16 and 17 depict a range of water costs, against volume × head product with insolation as a parameter.

X. FORMULAS USED FOR ESTIMATING THE SIZE OF A PHOTOVOLTAIC (PV) ARRAY IN A WATER PUMPING APPLICATION

A PV array is rated by its electrical power output at a temperature of 25°C under a solar irradiance of 1000 W/m^2 or:

$$W_p = n_{pv} \cdot A \cdot 1000 \tag{13}$$

where W_p is the array rating in peak watts (i.e., at solar noon), n_{pv} is the array conversion efficiency at a reference operating temperature, i.e., 25°C, and A is the cell/array area in square meters (*Note:* This area is total active cell/array area minus gridding metallization

---

* m^4 = Volume × head product.

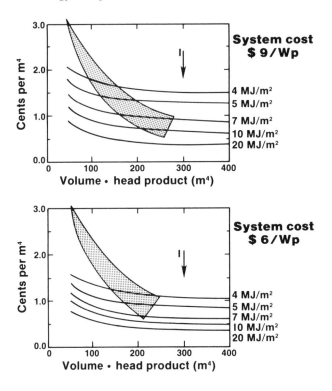

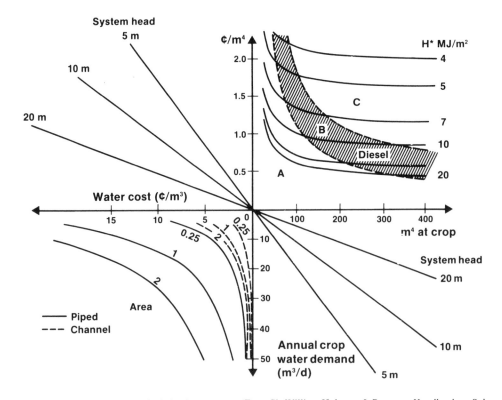

FIGURE 16.   Comparison of water costs between solar and diesel (the shaded area is a typical range of diesel pumping costs).

FIGURE 17.   Unit water costs for irrigation systems. (From Sir William Holcrow & Partners, *Handbook on Solar Water Pumping,* Intermediate Technology Power, UNDP Project GLO/80/003, February 1984. With permission.)

and framework). The active cell/array area required to provide a daily energy output of E,MJ for a daily solar irradiation (MJ/m^2) is given by:

$$A = E/(n_{pv} \times H) \tag{14}$$

where H is the daily irradiance in mega Joules per square meter.

Actual array efficiency must account for temperature and array electrical mismatch effects, i.e., operating off maximum power point position, thus:

$$n_{pv(e)} = F_m (1 - (T_{cell} - 25)n_r \tag{15}$$

The first term on the right of Equation 15, F_m, is the power mismatch factor or the ratio of electrical output under actual operating conditions to the output if the array were operated at its maximum power point assuming no maximum power tracking system is used. The second term is the reduction in efficiency due to the cells operating at a higher temperature than reference or standard conditions. T_{cell} is the temperature coefficient of the cell material.

Substituting for $n_{pv(e)}$ (effective array efficiency) from Equation 15, Equation 14 becomes:

$$A = E/(F_m (1 - (T_{cell} - 25)n_r H) \tag{16}$$

where n_r is the efficiency under reference conditions, i.e., 25°C at 1000 W/m^2 irradiance incident. A is the cell area in square meters required to provide a daily energy output of E, MJ. To determine the required array rating, Equation 16 is substituted into Equation 13:

$$W_p = 1000 \ E/(F_m(1 - (T_{cell} - 25)) \times H \tag{17}$$

using typical parameters: $F_m = 0.9$, $T_{cell} = 40°C$, and $H = 0.5\%$ per degree Celsius to give:

$$W_p = 1200 \times E/H \tag{18}$$

Figure 18 shows a general schematic of a PV pumping system to which the foregoing analysis would be applicable to.

XI. IRRIGATION WATER REQUIREMENTS

The quantity of water needed to irrigate a given land area depends on a number of factors, the most important being:

- Nature of the crop
- Crop growth cycle
- Climatic conditions
- Type and condition of soil
- Land topography
- Field application efficiency
- Transportation efficiency
- Water quality

Many of these vary with the seasons, and the quality of water required is never constant. The design of a small irrigation pump installation will need to take all these factors into consideration.

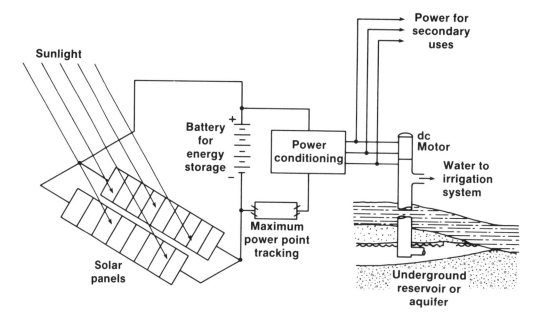

FIGURE 18.   Solar photovoltaic underground water system diagram. (From Sir William Holcrow & Partners, *Handbook on Solar Water Pumping*, Intermediate Technology Power, UNDP Project GLO/80/003, February 1984. With permission.)

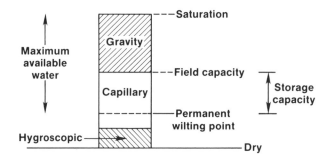

FIGURE 19.   Soil moisture quantities. (From Sir William Holcrow & Partners, *Handbook on Solar Water Pumping*, Intermediate Technology Power, UNDP Project GLO/80/003, February 1984. With permission.)

The crop takes its water requirements from moisture held in the soil. Useful water for the vegetation varies between two levels: the permanent wilting point and its field capacity (Figure 19). Water held by the soil between these two levels acts as a store. When this store approaches its lowest level, the crop will die unless additional water is supplied.

The rate of crop growth depends on the moisture content of the soil. There is an optimum growth rate condition in which the soil water content lies at a point somewhere between the field capacity and the permanent wilting point as shown in Figure 20. However, this point varies for different crops and for different stages of growth and so it is not easy to adjust the watering intervals and flow rates so that there is optimum growth. Multiple crops merely further complicate the situation.

An estimation of the quantity of water that may be required for irrigation can usually be obtained from agricultural experts and the United Nations Food and Agricultural Organization. The procedure involves these steps:

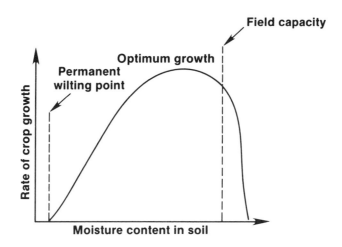

FIGURE 20. Rate of crop growth as a function of soil moisture content.
(From Sir William Holcrow & Partners, *Handbook on Solar Water Pump-ing*, Intermediate Technology Power, UNDP Project GLO/80/003, Feb-ruary 1984. With permission.)

1. Prediction methods are used to estimate crop water requirements, because of the difficulty of obtaining accurate field measurements.
2. The effective rainfall and groundwater contributions to the crops are subtracted from the crop water requirements to give the net required water.

XII. DYNAMICS OF COUPLING A DC MOTOR TO A PHOTOVOLTAIC (PV) ARRAY

A photovoltaic DC drive system (e.g., a water pumping and desalting system) is composed of three main components: the electric power source which in this discussion is a solar cell array, an electromechanical converter which is an electric motor, and a mechanical load formed by the pump(s). Each of these three components can be characterized by its operating plane: the electric plane I.V., the electromechanical plane (n,I) and the mechanical plane (n,T), respectively, as shown in Figure 21. In addition, there exists an optimum operating load line for each component which defines its optimal operating efficiency.

For a conventional (constant current) power supply we are interested in the efficient operation of the motor-mechanical load couple, and matching is made for these two com-ponents. For the photovoltaic system, we are interested in not only the maximum efficiency of the PV array, but also for minimum array cost. In other words, the overall optimal operation of the pumping/transport/desalting system is achieved if (1) the transformed me-chanical load converted by the electric motor matches the maximum power locus of the solar generator for differing levels of irradiance of the sun and (2) the transformed maximum power locus of the solar generator converted by the electric motor matches the maximum output of the mechanical load. If there exists an optimal line which satisfies both of these conditions, then the next parameter to examine is the stability of the mechanical load line. This will involve an analysis of its behavior under unprimed conditions and low levels of irradiance, among other things. It should be mentioned that the three components we are dealing with (array, motor, and pump) are individual devices and not necessarily compatible with each other in tandem. Therefore, we should seek a practical optimal system design which includes the choice of electric motor and mechanical load (pump type) on the one hand, and the motor control means and solar array generator interconnections on the other. In addition, the best system design should take into consideration the proportional efficiency, and efficiency variation, and the cost of each component.

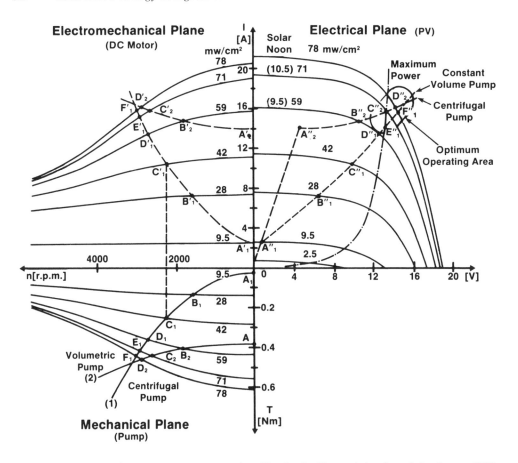

FIGURE 21. Motor-solar generator system operation. (Reprinted with permission from *Solar Energy,* 27(5), Appelbaum, J. A., Performance analysis of DC Motor-Photovoltaic Converter System II. Copyright 1981, Pergamon Press, Ltd.)

The practical optimum design should be represented by a transformed mechanical load, in the (I,V) plane that is as close as possible to the maximum power locus line of the solar array. In the (n,T) it should be as close as possible to the maximum power output of the mechanical load.[8,9] It should be noted that the transformed mechanical load is independent of the maximum power line and vice versa, i.e., it is effectively buffered.

The analysis of the system operation is best done in the operating plane of each component. The operating point is determined by the intersection of the component characteristics, and since the solar radiation varies during the day, the operating points form a load line. This line can be altered in different ways to achieve best performance. In addition, the analysis enables some insight of the systems operation, stability, and utilization under static conditions.

A graphic-static analysis, shown in Figure 21, can give some insight into the behavior of the system under transient operating conditions. Under these conditions a computer model of the system can be used effectively. Such computer analysis can be particularly useful in discerning stability criteria for intermittent illumination conditions, like scudding cloud passage over the PV array. A dynamic analysis at this level will give some indication of system damping, and its bearing upon system life of rotary bearings and shear forces on shafts, etc.

XIII. WATER DESALINATION

There are two basic methods of water desalting: distillation processes and membrane

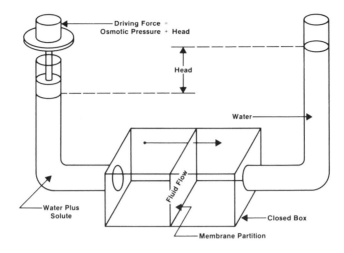

FIGURE 22. Reverse osmosis.

processes. In 1975, of the 346 desalting plants in use throughout the U.S. with capacities greater than 25,000 gal/day, about 0.5% utilized distillation processes with primarily nonelectrical energy sources. The remaining desalting plants utilized the membrane processes. Here the energy was almost entirely electrical. The two most common membrane techniques used are reverse osmosis (RO) and electrodialysis (ED). Systems in 1975 were producing 30.1 million gal/day. It is projected to reach 1295 million gal/day by the year 2000. The demand for desalted water is distributed among users such as municipal, industrial, electric power, tourism, and military. Municipal and industrial applications constitute the major users.

Both RO and ED employ semipermeable membranes to achieve purification of brackish and salt water. In ED, direct current electricity causes migration of ions through ion-permeable membranes to form concentrated and dilute solutions on opposite sides of the membranes. In RO, the salt components are filtered out as the water passes through the membranes under hydraulic pressure (see Figures 22 and 23). Electricity is used in RO systems solely for transporting water through various stages of the processes.

Figure 24 shows the flow diagram of a typical RO system. Feedwater from wells is chemically pretreated prior to desalting in order to minimize corrosion and fouling of membranes due to impurities such as iron, magnesium, or calcium carbonate. The water is then pressurized by high-pressure feed pumps to the operating range of 530 to 550 psig before it enters the RO modules contained in pressure vessels. The product water is piped to a reservoir for storage. The electrically driven pumps in the system are the high-pressure pumps, product pumps, well supply, and chemical pumps.

The performance on an RO or ED system can be described in terms of the recovery ratio and the energy required to remove given amounts of total dissolved solids (TDS) in the solution. The recovery ratio is defined as the quantity of product water as a percentage of feedwater and it depends strongly on the total dissolved solids (TDS) concentration. For an RO plant to produce potable water (TDS = 500 ppm), the recovery ratio can reach 90% for brackish water containing 1000 to 10,000 ppm of TDS, and 30% for seawater.

An ED plant also consists of a pretreatment section prior to the desalting process. The pretreatment section is utilized to provide removal of carbon dioxide and iron, partial brine softening, and clarification. As in the RO system, it is designed to reduce the majority of the TDS in solution before desalting in order to prolong the useful life of the membranes. Depending on the amount of TDS in the feedwater, the system can be designed to produce water of any desired quality by staging several membrane stacks in series. The ED systems,

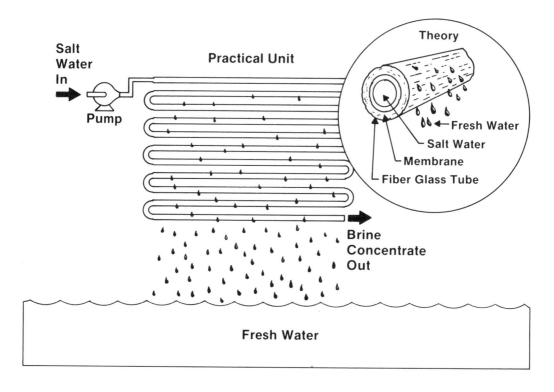

FIGURE 23.   Illustration of tubular reverse osmosis.

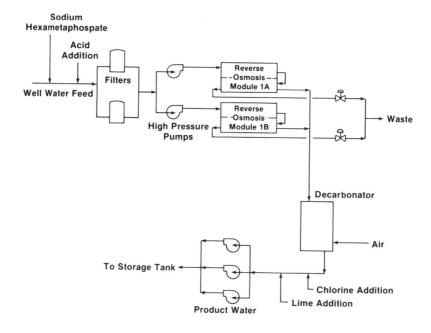

FIGURE 24.   Reverse osmosis plant process flow diagram.

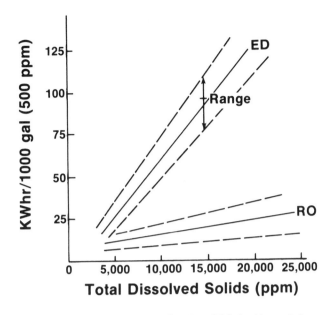

FIGURE 25.   Energy requirements for ED and RO desalting techniques.

in general, have a higher recovery ratio than the RO systems, as the brine solution is recirculated at the end of the process and mixed with incoming feedwater for further purification.

DC power that is supplied across the membrane stack varies according to plant sizes and the TDS concentration levels in the feedwater. The maximum DC sources have been about 600 V, 60 to 70 A for large plants, and they can range down to 15 V, at a few amps systems for very small plants. The voltage is adjusted to provide specific operating conditions. The energy requirements in an ED or RO plant can be specified in terms of kilowatt hours per 1000 ppm TDS removed per 1000 gal product water. Figure 25 shows the comparison of energy requirements to desalt water of increasing TDS for ED and RO systems. In both cases, as the TDS concentration increases and the recovery ratio decreases, higher feedwater flow and thus higher pump power is required to produce purified water at the constant rate. In addition, higher DC electrical potential across the membrane stacks is also required in the ED systems and, due to increased osmotic pressure, higher pressure feed pump is needed in the RO system. The energy requirements in a typical RO plant can be determined by:

$$\text{kWh/1000 gal product water} = \frac{7.28 \times 10^{-3} \times \text{pressure}}{\text{pump and motor efficiency} \times \text{recovery ratio}} \quad (19)$$

where 7.28×10^{-3} is a membrane flux term. In general, for every 1000 gal of potable water produced, the energy requirements are about 10 kWh.

In an ED system, the energy requirements are a function of low-pressure pumps (80 to 100 psi) and the electric potential across the stacks. During typical plant operation, the pump power totals to about 3 to 4 kWh/1000 gal of water produced and the DC power requirements for the stacks are about 4 to 6 kWh/1000 gal/1000 ppm of salt removed.

In both systems, water can be processed at slightly higher temperatures for better operating efficiency. In an ED plan, the membrane stacks will operate at maximum efficiency when the water temperature is at 43.3°C. However, when the temperature is above 43.3°C or below 21.1°C, the system performance begins to deteriorate. As for RO systems, the membranes can accept water temperature up to around 40.6°C. Direct current can be used as a power source for both RO and ED systems and the higher-and low-pressure feed pumps can be driven by DC motors. Normally, power is purchased from utilities in 480-V, 3-phase, 60-cycle AC, when and where available. In ED plants, AC is transformed by rectifiers and supplied across the membrane stacks.[10] Usually, the energy is supplied 24 hr/day to large desalting plants for continuous operation. However, as the plant sizes decrease, the systems are designed, utilizing water storage reservoirs, to operate on an intermittent schedule. The storage is sized to match peak load and cover low demand periods during plant shutdown.

In view of the above, it appears that photovoltaic energy may be a viable power source for both electrolysis and RO processes. Several favorable factors that result in such a conclusion are[11]

1.   DC can be used for both RO and ED plants. DC is a process requirement for ED plants. Drive motors for both ED and RO plants, while nominally AC, are convertible to DC.
2.   The systems can operate in an interruptible mode without serious start-up problems. Thus, with a properly sized storage medium, the PV-driven desalting system can be stand-alone in remote areas where utility grid power is not available.
3.   A large number of RO and ED plants are located in sunny desert areas where solar irradiance is high.

The most cost-effective application of RO on groundwater supply is seen to be as follows:

1.   Desalination of brackish borewater up to 7000 or 8000 ppm salinity, particularly water with relatively low silica, Fe, Mn, and Al levels.
2.   Small-scale desalination of surface water or potable water to produce a very pure product (for large-scale applications, ion exchange might be more popular).
3.   Seawater desalination in regions with either cheap electrical power or where space considerations are predominant (e.g., on offshore oil and gas drilling platforms).

In most areas with very brackish water (7000 to 35,000 + ppm), it is felt that waste heat processes will generally produce cheaper water; if waste heat is not available, then RO could still be the best choice. Solar-powered RO is possible and the most often quoted system, being PV cells plus RO. This has been shown to be attractive relative to other solar alternatives for most salinities. More recent work, particularly on high-temperature membranes, has increased the range of ED processes and improvements in the economics of ED desalination of seawater have been reported. The cost of desalting seawater by ED, however, is still considerably higher than either distillation or RO.

XIV. STORAGE OF WATER

One of the prime problems with solar energy is that it is not available on demand. For irrigation applications it may be critical that water be available to prevent a crop from drying out, and it is usually equally important to have water on demand for rural water supply. Thus, when using solar energy for water supply, the problem of storage must be dealt with. For irrigation, two types of water storage methods can be considered:

1.   Long-term storage in which water is stored from month to month to even out the demand pattern. This type of storage will permit irrigation on demand, and minimize the effect of variations on periodic water requirements. Once a consumption pattern has been established, a reservoir can be maintained that will cater for poor weather periods and times of high peak water demand.

2. Short-term storage which allows water to be stored from one month to the next month only. This serves two purposes: improved water management control and smoothing out day-to-day variations. For instance, on a day with high insolation values, a given proportion of the pumped product that is in excess of requirements can be set aside in storage.

Protocols of use, storage, and pumping can then be evolved that take into account crop needs, weather conditions, insolation values, etc. Such procedures are similar to those used when electrical battery storage methods are used with photovoltaic arrays.

Long-term storage for irrigation is generally not feasible due to economic and practical reasons. Small land holders do not usually have the space or the facilities to maintain large storage pools. Very often the soil itself can act as short-term reservoirs, depending upon soil porosity and evaporation rates. For rural water supplies other rules apply. It is essential to include a storage medium of some sort. This should meet several days demand during poor weather and low insolation values.

Storage should not add excessive head to the system and this parameter is a trade-off between evaporation rates. Where evaporation rates are high a cover might be added to the storage pool, or the surface treated with surfactants. In any case, storage tanks should be lined to minimize leakage or seepage.

XV. GROUNDWATER WELLS

A. Site Selection
Foremost among the factors which must be considered in selection of a site for a proposed well are local regulations governing proximity to such potential contaminants as cesspools, privies, animal pens, and abandoned wells. Also to be considered is the effect on adjacent wells, because if wells are located too close together excessive head losses may result.

B. Well Stations
A typical well station generally includes a small building for housing the pump, pump controls, metering, surge-control facilities, and chemical feed equipment. Submersible pumps do not require a pump house for protection, but if pump controls or chemical feed equipment are provided, an enclosure of some type is required. Since wellwater supplies are often pumped directly into the distribution system, a differential producer is usually installed for metering purposes. Moreover, chemicals may be added at the well station to minimize corrosion, control bacteria, decrease hardness, and inject fluorides into the water supply. Surges are usually controlled by installation of a surge valve in the pump discharge line. Controls may also be installed to permit starting and stopping the pump from remote central locations and to provide for measurement and control of well drawdown.

XVI. REVERSE OSMOSIS

Osmosis is the process in which a solvent flows across a membrane separating a stronger from a weaker solution. The solvent flows in the direction that will reduce the concentration of the stronger liquid. The flow of solvent between solution compartments can be observed as the liquid in the compartment of the stronger solution increases in volume. If these compartments are fitted with standpipes, the flow will continue until the level in the stronger solution compartment is higher than in the weak solution by a dimension which equals the osmotic pressure, as shown in Figure 22.

In reverse osmosis (RO), a driving force, the differential pressure across the membrane, causes water to flow from the stronger solution to the weaker. Therefore, the pressure

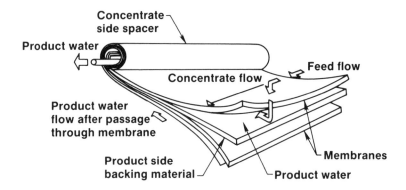

FIGURE 26.   Membrane fabrication.

required must exceed the osmotic pressure. This differential pressure is often greater than 300 psig (21 kg/cm^2) depending upon concentration differences. It averages about 10 psi for each 1000 mg/ℓ difference in total dissolved solids (TDS). The differences are influenced by the types of membranes used. In RO transport of water through the membrane is not the result of flow through definitive pores, at least not pores as commonly conceived. It is the result of diffusion, one molecule at a time, through vacancies in the molecular structure of the membrane material. The vacancies in amorphous polymers are in a state of flux, or are not fixed, while in crystalline material the vacancies are voids in lattice structures and are essentially fixed in number, position, and size. RO membranes are made of amorphous polymers, but usually contain some crystalline or less amorphous regions. RO has also been called hyperfiltration, indicating its relationship to a high-pressure filtration process. However, it should not be confused with utlrafiltration, which uses lower pressure and different membranes, because ultrafiltration removes essentially no solutes of low to intermediate molecular weight, while RO can remove even low molecular weight ionic species.

The salt removal or rejection obtained is primarily a characteristic of the semipermeable membrane. A given membrane may reject one ion more effectively than another; this rejection may be affected by the presence of other ionic species in the matter being processed. In general, the greater the water flux through a given type of membrane, the lower the salt rejection — the less salt retained in the concentrate and the more contaminated the product. Conversely, the higher the salt rejection rates, the lower the water flow through the membrane using the same applied pressure. These properties can be varied by changing the type of polymers used and methods of manufacturing and processing the membranes. As in electrodialysis (ED), potential for concentration polarization, scaling, and fouling are all serious considerations in designing process units to take advantage of the various membrane materials available.

A. Typical RO Designs

Membranes are formed into tubes (Figure 26) rolled from sheets and made into fine hollow fibers (Figures 27 and 28). To meet the required rejection rate, the design must maximize the flux rates per unit volume and minimize the problems from concentration polarization without seriously increasing the energy needed to operate the system.

It is extremely difficult to clean membranes once they are fouled. Pressure cycles help, as do some cleaning agents. Preconditioning water to be processed is almost always required. A Fouling Index has been developed as guidance in pretreatment processes for these tend to be expensive. The Fouling Index is based on the rapidity of plugging of a standard membrane filter. The test water is filtered at constant pressure and the time required to

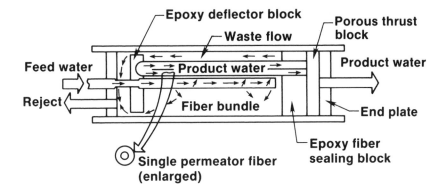

FIGURE 27. Membrane/fiber system.

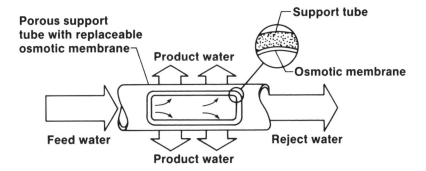

FIGURE 28. Membrane action.

produce 100 mℓ of filtrate is determined initially (t_0) and after 15 min (t_{15}). The Fouling Index (FI) is then determined from the relationship:

$$FI = (6.7 \times t_{15} - t_0/t_{15}) \tag{20}$$

To avoid fouling RO membranes, the FI of the feed water should be less than 3. Chemical treatment of polyphosphates is also used to attempt to reduce plugging of RO membranes with precipitates of $CaCO_3$, $CaSO_4$, and $Mg(OH)_2$.

B. Demonstration[11]

A generalized schematic of an RO plant with pretreatment is shown in Figure 24. RO elements and the extent of pretreatment are determined by site and raw water conditions.[12] As an example of what can be done with solar photovoltaics by way of groundwater desalination, Figure 29 shows a schematic of a desalination plant built by Mobil-Tyco at Jiddah, Saudi Arabia on the eastern shore of the Red Sea. Electrical power for pumping and RO hydraulic pressure is supplied by an 8-kW peak nominally rated PV array, under 100 mW/cm² AM 1.0 insolation at 25°C. This derates to approximately 6 kW after temperature rise, lowered irradiance, and dust buildup effects are taken into account. This 6 kW produces, dependent upon the season, approximately 46 kWh at 330 V DC.

After energy for auxiliaries and battery charging is subtracted, 34 kWh produces 480 gal of water at a salinity of below 500 ppm, in conformity with World Health Organization standards. Feedwater salinity is 43,000 ppm taken from a 3-m well by a shallow-well jet pump. Apparently a jet pump was used for its reliability and long life pumping very corrosive feedwater. This installation consists of a brushed DC motor rated at 1 hp (1750 rpm at 240 V) to provide RO hydraulic pressure. Product water flow rate yield is 1.15 gal/min and the membranes at the hydraulic pressure used have a 22% recovery rate (see Figure 27), thus requiring 5.2 gal/min of feedwater to be pumped from the well. Thus, this prototype PV water pumping and desalting system produces potable water at the rate of 18 gal/kWh of electricity produced. This gives, on the late 1990s projected cost of photovoltaic systems (including both array and balance-of-system), a cost of 15¢/kWh or 0.9¢/gal. This is expensive water in contrast to well-watered areas, but in places like Saudi Arabia it might well be considered cheap.

XVII. PERMANENT MAGNET BRUSHLESS DC MOTORS

As has already been noted, the need for reliability in remote locations is paramount. To this end a reduction in component count is always to be desired. In pumping and desalting groundwater one of the most vulnerable elements is the motor(s) that provide hydraulic pressure and their associated DC to AC inverters.

The brushless DC motor concept is not a new one. The first such motor was built commercially about 25 years ago. At that time the cost of electronic commutation was high and brushless DC motors were too expensive to be popular.[13] The period since then has seen the advent of power electronics with relatively cheap high-powered switching semiconductors, and the development of rare earth and ceramic permanent magnets. Both of these occurrences have made possible the marketing of brushless DC motors both for safety attributes and inherent reliability factors. Conventional brushed-type permanent magnet DC motors are compact, reliable, and offer vastly improved controllability due to their linear speed-torque characteristic. However, the brushes in these motors eventually will wear out, and motor life, as a consequence, is inherently limited. Permanent magnet brushless motors, conversely, are by their nature, more reliable because they do not have brush-related problems. In some applications, like remote site deployments, brush problems can be a serious maintenance consideration. Furthermore, the electronic commutation of brushless DC motors offers greater flexibility with respect to direct interfacing with digital commands coupled to microprocessors. This attribute leads to considerable energy savings in pumping operations where variable voltage-variable frequency (V^3F) applications vary pump speed, instead of throttling by use of gate valves. These and other basic advantages of brushless motors over conventional brush-type motors are summarized as follows:

- Brush replacement and problems associated with it are nonexistent.
- Long life — with suitable bearings, lifetimes in excess of 20,000 hr are possible.
- Very little maintenance is required. There are no brush-related contamination problems.
- They are ideal for use in water immersion, where brushed types do not generally work well.
- Fire hazards are minimized for there are no sparking brushes.
- Electrical and acoustic noise is minimal.

Since the use of brushless motors is not widespread at this time, it is appropriate to discuss their principles of operation. Brushless DC motors differ from conventional DC motors in that they employ electrical rather than mechanical commutation of the field windings.

The configuration of the brushless DC motor most commonly used is one in which the rotor consists of permanent magnets and "black iron" support and whose commutate windings are located external to the rotating parts, as shown in Figure 30, where both the brushed and brushless types are depicted. Thus, the brushless type, compared to the brushed, is an inverted configuration. Although this configuration is the most common, the structure of

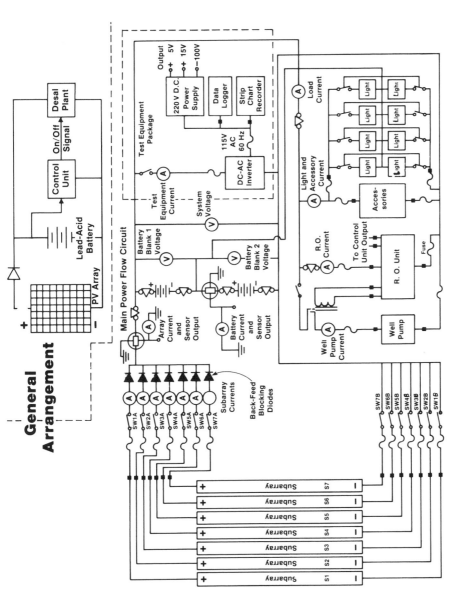

FIGURE 29. RO schematic circuit—general arrangement.

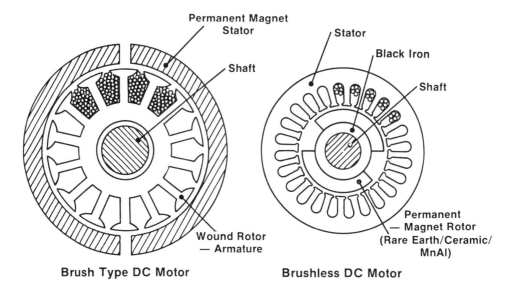

FIGURE 30.   Illustration of the difference in construction of a brushless DC motor and a conventional permanent magnet brush-type motor.

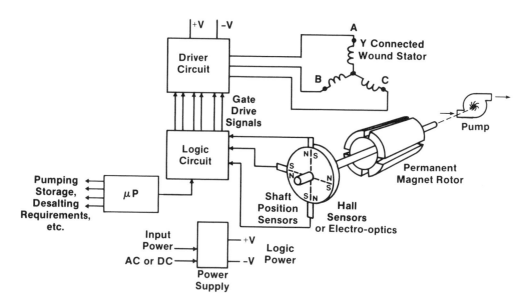

FIGURE 31.   Essential parts of a three-phase brushless DC motor.

the brushless motor can be arranged such that the inner member is the armature and the outer magnet member is the rotor.[14]

The block diagram shown in Figure 31 illustrates the essential parts of a brushless DC motor. The Hall effect or magnetic switches sense the angular position of the shaft and feed this information to the logic circuitry. The logic encodes this information and controls the six switches in the driver circuit. Appropriate windings, as determined by the rotor position, are sequentially excited with the correct polarity power. Consequently, the magnetic field generated by the rotor windings rotates in relation to the shaft position, reacts with the field of the rotor permanent magnets, and develops the required torque. This arrangement essentially simulates the function of a brush-type motor having a three bar commutator.[15,16]

XVIII. TYPES OF PUMPS FOR USE WITH SOLAR PHOTOVOLTAICS

There are a number of types of pumps that will work from depths of the order of 30 m. These are[17]

1. Jack pump
2. Jet pump
3. Centrifugal pump with DC motor on surface
4. Progressive cavity pump with DC motor on surface
5. Centrifugal pump with submersible DC or AC motors

We have determined that the centrifugal pump with the submersible motor (#5) is the optimum selection for efficiency, reliability, and economy. Discussion of the various pump types follows.

A. The Jack Pump[18]

Jack pumps are characterized by a cylinder at the bottom of the well below water level and a piston in the cylinder moved by a pump rod extending to the surface where "the jack" converts rotary motion to vertical reciprocating movement. There may or may not be a counterbalance to equate the weight of the pump load and column of water to be lifted on each up stroke. Even with a carefully adjusted counterbalance, experience has shown that currents to the motor vary by a factor of five between the up and down strokes. Unless batteries are used on the systems, this promotes extremely inefficient use of the photovoltaic energy source.

B. Jet Pump

Jet pumps have a jet set up as an aspirator below water level in the well. Water from a storage tank is pumped at high velocity down the well and up through the jet. The high velocity caused by the constriction in the jet causes a reduced pressure so that the static head of water above the jet forces additional water into the discharge pipe which is then raised by the pump pressure to the surface and discharged. The efficiency of this system is between 5 and 10% and certainly is not suitable for use with photovoltaics. Photovoltaic arrays are the most costly portion of a photovoltaic water pumping system and will continue to be so for the foreseeable future. It is therefore essential that the pump and motor be of the highest efficiency possible to minimize the size and concurrent cost of the photovoltaic portion of the system.

C. Centrifugal Pump with DC Motor on Surface

Pumps of this type are made by a French firm, Pompes Guinard, and have proven effective. However, they are expensive and have problems in that the shaft from the motor on the surface must run the whole depth of the well to the pump below the water level. There are water-lubricated bearings at 1-m intervals which require maintenance. In addition, the friction of the water around the shaft as well as the many bearings all increase losses. The main advantage of this configuration is that the surface-mounted DC motor is easily accessible for periodic replacement of brushes.

D. Progressive Cavity Pump with DC Motor on Surface

This pump is in many ways similar to the Guinard, and shares the same frictional losses. Its advantages are that the progressive cavity pump is a positive displacement pump and it will pump water irrespective of the speed of rotation. Normal speed would be of one half to one third that of the centrifugal pump, which would reduce friction losses in the shafting;

these are more than equated by the increased friction in the progressive cavity. Should the drive portion of the system fail, the pump can be turned by hand at a few revolutions per minute and will still pump water. The progressive cavity pump, even when driven by a DC motor, requires a high starting torque which in turn requires a high current, making it unsuitable for use with a photovoltaic array power source that does not include battery storage.

E. Centrifugal Pump with Submersible AC or DC Motors

Since submersible AC pumps have been in use for many years and have established their reliability with a large source availability at competitive prices, it is the prime choice for the pump to be used when there is a requirement for the absolute minimum of maintenance with low initial capital. A major problem with AC motors is the high starting current required. When an AC motor connected to a pump is started using power from a commercial network at 50 or 60 Hz, the starting current is of the order of five times the running current. This is obviously impractical if the power source is to be a photovoltaic array without batteries.

XIX. DETERMINATION OF PUMP CAPACITY

The capacity that can be obtained from any particular well is dependent on such factors as screen size, well development, permeability of the aquifer, recharge of the groundwater supply from rainfall and streams, and the head available. The basic procedure used in sizing a pump for well service involves drilling of the well and performing a test operation. The first step involves determination of the static head, or elevation of the groundwater table prior to pumping. Pumpage at various rates is then conducted and the drawdown associated with each pumping rate is determined. A plot of drawdown vs. pumping rate can then be derived. Pumping rate is usually measured by a weir, orifice, or pitot tube, and drawdown is determined with a detector line and gauge or with an electric sounder. From the test data and from a preliminary layout of discharge piping, a system head curve can be derived, with drawdown added to friction losses for each pumping rate. The pump characteristic curve can then be superimposed on the system head curve to determine the capacity that can be attained with a specific pump. It should be noted that pump curves for line shaft pumps are based on the results of shop tests, which do not allow for column friction or line shaft and thrust losses. Consequently, the laboratory characteristic curve for any line shaft pump must be adjusted to actual field conditions. Field pumping head can be determined by subtracting column friction losses from the laboratory head. Field brake horsepower is determined by adding shaft brake horsepower (which depends on shaft diameter and length, and on rotative speed) to laboratory brake horsepower; field efficiency is determined from the formula:

$$\text{Field efficiency} = \frac{\text{gpm} \times \text{field head}}{3960 \times \text{field brake horsepower}}$$

Since thrust loads cause additional losses in the motor bearing, it is necessary to determine the additional horsepower required to overcome thrust losses. Total thrust load is equal to the sum of the shaft weight and hydraulic thrust (which varies with laboratory head for any particular impeller), and losses due to thrust amount of approximately 0.0075 hp/100 rpm/1000 lb of thrust. Motor efficiency is then calculated by dividing the full load horsepower input (without thrust load) by the sum of full load horsepower input and loss due to thrust. Overall efficiency then equals the product of field efficiency and motor efficiency.

As a result of the efficiency losses produced by shaft weight and length in line shaft pumps, it is usually more economical to use a submersible pump at depths over about 500 ft. Sizing a submersible pump requires calculations similar to those for the line shaft pump. However, the submersible pump installation requires a check valve in the column pipe, which must be considered in the determination of friction losses. Moreover, the efficiency losses resulting from the motor cable (expressed as a percentage of input electrical horse-power) must be considered in determining overall efficiency, which can be calculated from the formula:

$$\text{Overall efficiency} = \frac{\text{water hp} \times (\% \text{ motor efficiency} - \% \text{ cable loss})}{\text{shop bhp} \times 100} \tag{21}$$

where

$$\text{water hp} = \frac{\text{gpm} \times \text{ field head}}{3960}$$

Cable size must be selected on the basis of motor horsepower and motor input amperes, voltage, and cable length.

XX. CONCLUSIONS

The use of alternative energy systems employing solid state technology for agricultural and other purposes will be increasingly evident in the future. The displacement of fossil fuels by this newer, environmentally benign technology will be evolutionary. As the costs of photovoltaics become competitive, there will be increasing incidence of use in remote areas for such duty as groundwater pumping and desalination.[19,20] With competitive front-end costs reduced, operation and maintenance will be minimal for this type of deployment, and, of course, fuel delivery to site will not be required.[21] The future use of thin-film and multijunction technologies will further reduce costs and increase the efficiencies of these truly remarkable devices.

REFERENCES

1. **Heimes, F. J. and Luckey, R. F.,** Method for Estimating Historical Irrigation from Ground Water in the High Plains in Parts of Colorado, Kansas, Nebraska, New Mexico, Oklahoma, South Dakota, Texas and Wyoming, Water Resources Div. Geological Survey, Denver, Colo., NTIS P.C. A04/MF A01, May 1982.
2. **Roger, J. A.,** Water and photovoltaics in developing countries, *Solar Cells,* 6, 295, 1982.
3. **Bird, R. E., Hulstrom, R. L., and Lewis, L. J.,** Terrestrial solar spectral data sets, *Solar Energy,* 30(6), 563, 1983.
4. **Solar Energy Research Institute,** *Solar Radiation Energy Resource Atlas of the U.S.,* SERI/SP-642-1037, October 1981.
5. **Rauschenbach, H. S.,** *Solar Cell Array Design Handbook,* Van Nostrand Reinhold, New York, 72, 1980.
6. **Longrigg, P.,** Rapid Charging of Lead-Acid Batteries for Electric Vehicle Propulsion and Solar Electric Storage, SERI/RR-742-1068, June 1981.
7. **Sir William Holcrow & Partners,** *Handbook on Solar Water Pumping,* Intermediate Technology Power, UNDP Project GLO/80/003, February 1984.
8. **Brunstein, A. and Kornfeld, A.,** Analysis of solar powered electric water pumps, *Solar Energy,* 27(3), 235, 1981.
9. **Appelbaum, J. A.,** Performance analysis of DC Motor-Photovoltaic Converter System II, *Solar Energy,* 27(5), 421, 1981.
10. **Gerofi, J. P. and Fenton, G. G.,** Comparison of solar RO and solar thermal desalination systems, *Desalination,* 39, 95, 1981.
11. **Wood, J. R. et al.,** A Stand-Alone Seawater Desalting System Powered by an 8 kW Ribbon Photovoltaic Array, IEEE Photovoltaic Specialists Conference, Orlando, Fla., 1981.

12. **Bushnak, A. A.,** Desalination by Reverse Osmosis, University of Riyadh, Saudi Arabia, Proc. Solar Desalination Workshop, Denver, Colo., SERI/CO-761-1077, March 1981.
13. **Woodbury, J. R.,** The design of brushless dc motor systems, *IEEE Trans. Ind. Elect. Cont. Inst.,* IECI-21, No. 2, 1974.
14. **Bishop, J. E.,** *Reviving direct current electric motor,* Wall Street Journal, Sept. 14, 1983, p. 29.
15. **Maslowski, E. A.,** Electrically Commutated DC Motors for Electric Vehicles, DOE/NASA 51044-14 NASA TM-81654.
16. **Vaidya, J. E.,** Optimization of brushless DC motor design, *Drives and Controls International,* 2, No. 5, July 1982.
17. **Waddington, D. and Herlevich, A.,** Evaluation of Pumps and Motors for Photovoltaic Water Pumping Systems, SERI/TR-214-1423, June 1982.
18. **Matlin, R. W. et al.,** PV water pumping with reciprocating volumetric (jack) pumps, 15th IEEE Photovoltaic Specialists Conf., May 1981, 1386.
19. **Smerdon, E. T. et al.,** *Energy in Irrigation in Developing Countries,* USAID, December 1980.
20. The Economics of Solar Water Pumping in Developing Countries, Strategies Unlimited, 1981.
21. *Small-Scale Solar-Powered Water Pumping Systems: The Technology: Its Economics and Advancement,* Sir William Halcrow & Partners, June 1983.

Chapter 3

WIND ENERGY APPLICATIONS IN AGRICULTURE*

R. N. Clark

TABLE OF CONTENTS

* Contribution from United States Department of Agriculture, Agriculture Research Service, Bushland, Texas.

I. INTRODUCTION

Wind power was first used to power small boats. As man learned to control these crafts, he built larger boats with larger sails. Gradually, sails were also placed on rotating booms, causing shafts to turn, and thus grinding wheels or pumps. Archeological records have shown that early forms of windmills were used by the Chinese, Egyptians, and Babylonians.

Written references to windmills are found in manuscripts of the early centuries A.D., with the Persian vertical-axis windmill described in some detail in the 7th century A.D. (Figure 1). Many European documents after the 13th century refer to various windmill designs. By the 1700s, two basic windmill designs had developed in Europe: the post mill and the tower mill.[1]

The post mill consisted of a large house and a tail pole (Figure 2). The tail pole was used to turn the structure and bring the turbine into the wind. These mills were primarily used along the coast to pump water and grind grain.

The tower mill was more like the modern windmill in that it consisted of a rotor and tail mounted on a fixed tower (Figure 3). Because of the ease of orientation, the tower mills were used in locations that did not have significant prevailing wind direction.

The rotor or sails were constructed primarily of wood, reeds, and canvas prior to 1900. Many different concepts and designs were used, including springs and shutters to increase or decrease the sail area. The sails on the larger English windmills were about 12 m long and 3 m wide. Peak power often reached 30 kW from these larger mills.[1]

Windmill development in the U.S. began by duplicating the European design, but these machines did not provide the flexibility needed to withstand the fickle weather of the Midwest. Daniel Holladay began making wind machines in 1857 that were self-regulating using paddle-shaped blades that would pivot, or feather, as the windspeed increased. After the Civil War, the Eclipse windmill was introduced (Figure 4). The Eclipse was the first to use a solid wheel assembly and a side vane to turn the rotor out of the wind as the velocity increased.[2] Both the Holladay and Eclipse used a reciprocating-type pump driven by a crank or offset cam. These systems worked well to lift water from the deep underground supplies of the West and Midwest. Windmills as large as 6 m in diameter were common along the railroads to provide water for the steam engines. Enclosed gears, metal wheels, and towers improved the systems until they operated well in light winds and ran smoothly. At the turn of the century, it is estimated that 200 companies offered windmills that were used to power saws and shell corn as well as pump water.

By 1930, wind power was also used to generate electricity. These electrical generating systems were quite different from the multiple-bladed water pumpers in that they usually had two or three blades rotating at a much higher speed. The electrical output was normally 12 or 24 V (DC), and incorporated batteries for energy storage. Jacobs Wind Electric Company reported selling tens of thousands of these units between 1931 and 1957.[3] Wind electric generators were removed with the installation of electric power lines by the Rural Electrification Administration (REA). However, windmills never entirely disappeared, with water pumping units remaining in the western U.S. to provide water for livestock in remote areas.

With the oil embargo of 1973 and the resulting desire to develop alternative energy sources, wind energy again became of interest. The U.S. Departments of Energy and Agriculture began to develop windmill designs based on new aerodynamic theories, and enthusiasts began to restore old wind electric plants. By 1984, some 8000 new units manufactured by over 60 companies were installed, having a combined capacity of almost 300 MW.[3] This rapid development illustrates the continued interest and feasibility of wind as an alternative energy source.

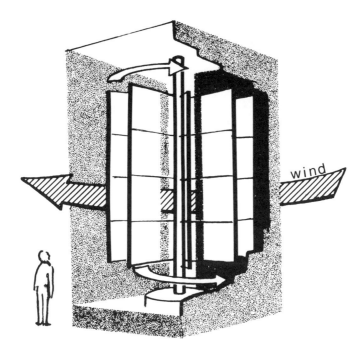

FIGURE 1.   Persian vertical-axis windmill used to grind grain in 700 A.D.

II. WIND ENERGY PRINCIPLES

The wind contains kinetic energy that is harvested by a rotor or wheel and transferred to a rotating shaft. Energy in the shaft is then used to directly pump water, drive an electric generator, or produce heat. Therefore, the energy in the wind is normally converted to either mechanical, electrical, or heat energy. Power is frequently used in describing the performance of wind machines and is a measure of the energy extracted during a specific period of time. The theoretical power in a wind stream is determined by

$$P = \frac{1}{2} D A V^3 \tag{1}$$

where P is the power in watts, D is the air density in kilograms per cubic meter, A is the cross-sectional area in square meters, and V is the windspeed in meters per second. Actual power harvested by a wind turbine is less than the theoretical because of power losses in the system. The efficiency or power coefficient (Cp) is determined by

$$Cp = \frac{\text{Power delivered}}{\frac{1}{2} D A V^3} \tag{2}$$

III. WIND CHARACTERISTICS

A. Windspeed
Wind is derived from temperature and pressure differences often caused by solar heating of the earth. Uneven heating of the surface of the earth caused by air, water, and land,

FIGURE 2.   The English post mill used to pump water in 1700.

create daily variations in wind movement. Also, because of yearly movement of the earth, seasonal variations are developed. To harness the wind, you must become familiar with its patterns and be able to select a suitable site. Wind is best described by Park[4] as "a fickle servant", and "It may not be available when you need it, and you can be overwhelmed by its abundance when you don't".

Historically, windspeeds have been measured near airports to assist aircraft in landing and takeoffs and near evaporation pans to give better indications of why evaporation rates vary. Agriculturalists have known the importance of windspeed measurements in predicting evapotranspiration rates, field drying rates, and loading on farm structures for many years. For all these applications, a simple daily wind run total or an instantaneous peak value was usually adequate to predict the expected outcome. However, because wind power is dependent on the cube of the windspeed, a windspeed distribution is needed. Figure 5 illustrates an actual windspeed distribution for Amarillo, Tex., showing the number of hours or percent

FIGURE 3. The tower mill used to pump water and grind grain in 1800.

of time that a given windspeed could be expected. Researchers have examined several mathematical distributions to determine which best represents actual windspeed distribution. The Rayleigh distribution was selected as providing the simplest, best approximation of windspeed characteristics nation-wide and is compared in Figure 5 to an actual distribution.

The Rayleigh windspeed distribution gives a ratio of time the wind blows within a given windspeed band, probability interval of V to V + dv, and the total time under consideration. The Rayleigh is a Weibull distribution with a constant of two and depends only on mean windspeed. It is defined by

$$F(V)dv = \frac{\pi}{2} \frac{V}{\overline{V}^2} \exp\left[-\frac{\pi}{4} \left(\frac{V}{\overline{V}} \right)^2 \right] dv \tag{3}$$

where F(V) is the Rayleigh frequency distribution as a function of V, $\overline{V}$ is the mean windspeed, V is the instantaneous windspeed, and dv is the windspeed probability interval.

Figure 6 shows a comparison of a measured wind power distribution with the Rayleigh distribution for Amarillo, Tex. Notice that the peak wind power is diminished by the Rayleigh prediction, but that the width is increased. In most cases, the Rayleigh distribution predicts the wind characteristic with less than a 10% error.[5]

FIGURE 4.   The Eclipse windmill used in the Great Plains for pumping water
beginning in 1870.

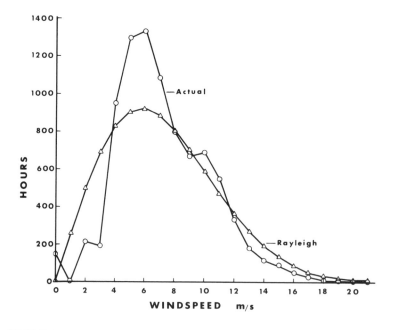

FIGURE 5.   Average windspeed distribution for Amarillo, Texas at a height of 12.1 m.

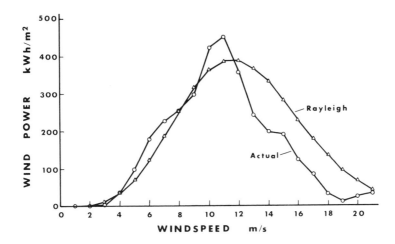

FIGURE 6. Average wind power distribution for Amarillo, Texas at a height of 12.1 m.

Because local winds are influenced by temperature and pressure differences, it is important to determine the prevailing wind direction. Often each season will have its own prevailing wind direction, and in mountain valleys there is usually a diurnal change in wind direction. Wind direction is normally up the mountain from noon to sunset because of the daytime heating, and then reverses direction and flows downward from evening to sunrise.[6] Wind direction determinations are much easier in noncomplex terrain, but are important when selecting a site for your wind turbine.

B. Wind Shear

As the wind passes over the surface of the earth, wind close to the surface is retarded due to the roughness caused by trees, buildings, and crops. The extent to which windspeed is decreased near the surface of the earth and the resulting variation with height is called "wind shear". The height at which free wind (100% of potential) is available depends on the surface terrain (Figure 7). The free wind over smooth terrain (water) could be as low as 40 m, but when determined over complex terrain (large cities), the height could be as high as 500 m. The extent of wind shear is also influenced by temperature and humidity. Cold, damp air has a larger shear than warm, dry air.

Meteorologists have suggested several equations for adjusting the windspeed measured at one height to represent the windspeed at another height. The most common equation is

$$V_2 = V_1 \left[\frac{h_2}{h_1}\right]^N \qquad (4)$$

where V_2 is the windspeed at the new height (h_2) and V_1 is the known windspeed at height (h_1). Detail measurements have shown that this equation does not always predict the correct windspeed at the new height because of variations in moisture, pressure, and temperature profiles as well as differing wind profiles. It is intended to represent average or long-term conditions.

C. Windspeed Measurements

The most common way of determining the windspeed is to use an instrument called an anemometer to measure the windspeed. The most popular type of anemometer is the cup-type, consisting of small cups mounted on a rotating shaft. Many are a small DC generator to produce a direct, readable voltage that is proportional to windspeed. The physical size

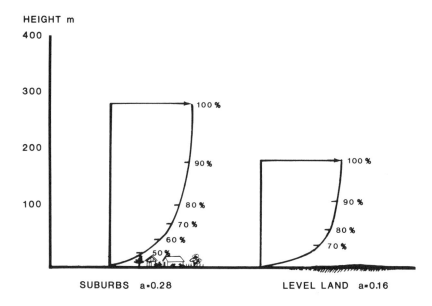

FIGURE 7.   Windspeed profiles over flat, open country and urban areas or large farmsteads.

and configuration of anemometers have been as varied as the manufacturers, with little standardization.

Anemometers used to measure windspeeds near evaporation pans have usually been at a height of 0.5 m and have been of the total wind run type, indicating the number of kilometers of wind passing the anemometer per day. Other wind measurements for predicting evapotranspiration have usually been taken at 2 m, with few measurements made above 3 m. Anemometers at airports have been placed at the most convenient places, like on top of buildings or along the runway. The heights have not been standardized and many different heights are found for the same airport in the records. Standard anemometer heights have been used more in the areas of environmental quality measurements than in any other area. Standard anemometer heights for nuclear and coal-fired electrical generating plants are 10 and 40 m, respectively.

Therefore, before using reported windspeed data, a determination of how data was collected (recording, frequency, averaging time, etc.) is needed along with the height of the anemometer. Many studies have been made using recorded wind data from the National Weather Service, USDA, and other government and private agencies. Figure 8 is the result of a study by the U.S. Department of Energy.[7] This map presents wind data as wind power classes rather than actual windspeeds. As a part of this study, 12 regional maps were prepared for the U.S. that provide sufficient detail to make a preliminary analysis of the wind energy potential for a particular site.[8]

Because of the complex nature of the wind flow patterns at any site, it is recommended that actual windspeed data be collected for at least 3 months for each site. The anemometer should be placed at the hub height of the planned machine to eliminate inherent errors in predicting windspeed at different heights. The anemometer should also be capable of supplying data to a recorder at least every 5 min. It is important to determine the variation in the windspeed because it affects the energy produced. Many states provide loan anemometers for making individual site analysis.

IV. SITE ANALYSIS AND SELECTION

A. Windspeed Distribution

The most important decision concerning the use of a wind energy conversion system is

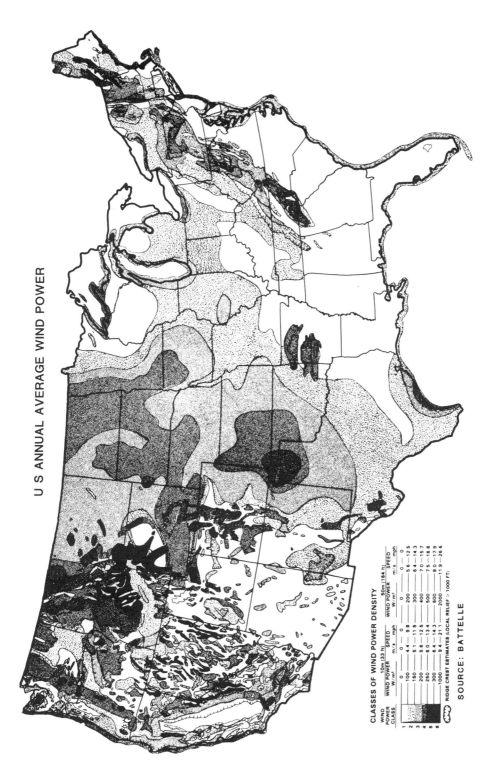

U S ANNUAL AVERAGE WIND POWER

CLASSES OF WIND POWER DENSITY

WIND POWER CLASS	10m (33 ft) WIND POWER W/m²	10m (33 ft) SPEED m/s	10m (33 ft) SPEED mph	50m (164 ft) WIND POWER W/m²	50m (164 ft) SPEED m/s	50m (164 ft) SPEED mph
	0	0	0	0	0	0
1	100	4.4	9.8	200	5.6	12.5
2	150	5.1	11.5	300	6.4	14.3
3	200	5.6	12.5	400	7.0	15.7
4	250	6.0	13.4	500	7.5	16.8
5	300	6.4	14.3	600	8.0	17.9
6	1000	9.4	21.1	2000	11.9	26.6

~~~ RIDGE CREST ESTIMATES (LOCAL RELIEF > 1000 FT)

SOURCE: BATTELLE

FIGURE 8. Estimates of average annual wind power for the continental U.S.

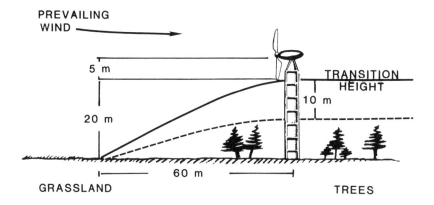

FIGURE 9.   The effect of trees and buildings on the transition and equilibrium height.

the selection of the site. Because of the high initial cost of wind systems, it is imperative that the best site be used rather than just a good one. Three main questions should be answered when doing a site analysis:

1.   Is there sufficient wind for the machine to produce usable power at least 50% of the time?
2.   What effects will surface terrain have on the wind profile?
3.   What barriers might affect the free flow of the wind?

The speed and duration of wind is determined from the windspeed distribution which can be calculated as described in Section III or by actual measurements. It is important that an analytical determination be made because most people tend to overestimate their wind resource. Only sites that have an annual average windspeed in excess of 5 m/sec at a 10-m height should be considered for wind power.

B. Surface Roughness

The roughness of the terrain determines the amount of windspeed reduction that occurs near the surface. Selecting a good site in flat terrain is much easier than in hilly or mountainous terrain. When selecting a site in flat terrain, changes in terrain often cause a transition in the height of the retarded wind flow region. Figure 9 shows the effect of trees on the transition height and its downwind result. Other large natural changes in terrain have similar effects and should be considered in the site selection process. Oftentimes, an increase in tower height will provide the needed change to place the rotor in the free wind, rather than moving the entire unit to another location. Performance data have shown that machines on 25-m towers do not perform as well as machines on 30-m towers in Iowa, while little difference in performance is noted between the two in west Texas.[9]

Determining the effect of surface terrain in hilly or complex terrain requires much work and analysis. Wind profiles over hills show that areas of acceleration occur as well as areas of turbulence and decreased flow (Figure 10). The shape of the hill determines the magnitude of each of the phenomena. Normally, the top and the upper half of hills on the windward face or where prevailing wind is tangent to the site are considered good sites because these are points of wind acceleration. The leeward sides and lower portions of the hills should be avoided because of reduced speeds and turbulence. Also, larger wind turbines should be placed where the entire rotor sees wind of a similar characteristic to avoid uneven loading on the rotor. Flow through valleys, canyons, basins, and gaps each create a special situation that requires extra evaluations when selecting a site.[6]

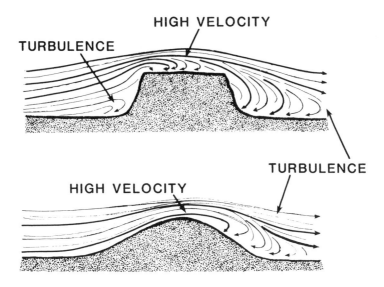

FIGURE 10. Schematic of wind flow patterns over hills showing areas of turbulence.

When first evaluating a potential site, it is useful to observe the vegetation. Vegetation is often deformed by high average winds, and can be very helpful in remote areas to evaluate the site prior to installing anemometry. Figure 11 shows a rating scale developed by Griggs-Putnam[10] for estimating the windspeed. Four easily observed deformities of trees are listed and defined:

- *Brushing* — Branches and twigs bend downwind like the hair of a pelt that has been brushed in one direction only. This deformity can be observed in deciduous trees after leaves have fallen. It is the most sensitive indicator of light winds.
- *Flagging* — Branches stream downwind, and the upwind branches are short or have been stripped away.
- *Throwing* — A tree is wind thrown when the main trunk and the branches bend away from the prevailing wind.
- *Carpeting* — This deformity occurs because the winds are so strong that every twig reaching more than several inches above the ground is killed, allowing the carpet to extend far downwind.

Using Figure 11 and the description above, mean annual windspeeds can be estimated from Table 1.

C. Man-Made Barriers

Even though the terrain at a particular site may be ideal, man-made structures such as houses, barns, etc. can create a barrier which severely reduces windspeeds. Buildings and trees hinder wind flow and not only cause lower windspeeds in their wake, but also create turbulence. The reduced windspeed in the vicinity of such obstructions will affect the power that can be produced by a wind generator. Also, the turbulence caused by the wind swirling about obstructions will shorten the lifespan of the unit. Small wind turbines, however, are best located near the point where the energy is to be used since the cost of transmission lines from a good site 1 km away may exceed the cumulative value of the conventional energy displaced.

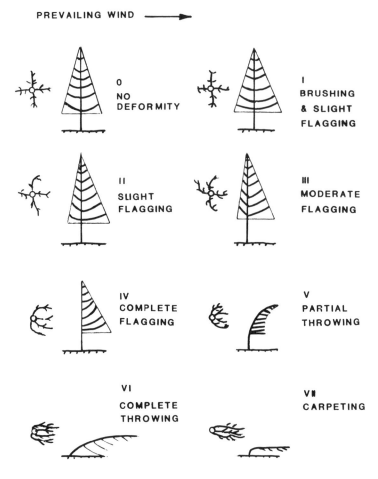

FIGURE 11.   Windspeed rating scale based on the shape of the crown and degree twigs, branches, and trunk are bent.

Table 1
MEAN ANNUAL WINDSPEED VS. THE GRIGGS-PUTNAM INDEX

	Griggs-Putnam Index (Figure 11)					
	I	II	III	IV	V	VI
Mean annual windspeed (m/sec)	3—4	4—5	5—6	6—7	7—8	8—9

A site must be far enough upwind or downwind to avoid the disturbed wind flow around an obstruction. Buildings produce different effects on wind flow than do trees and bushes. The turbulent zone over buildings and trees is shown in Figure 12. Turbulence is the result of many factors, but is primarily due to the height of the building or tree (H). Note that even the flow of air downwind from the building is affected. A machine should be located away from the building by a distance of 2H upwind and 10H downwind.

Shelterbelts or windbreaks cause an even greater effect on wind flow than do buildings. The kinds of trees in the windbreak, spacing, and height are the controlling factors. For a tower equal in height to that of the trees, power in the wind is reduced 80% at a distance of 10H downwind from the windbreak. A general rule of thumb is to erect a tower tall enough to clear the highest obstruction by 8 m and to site it at least 90 m from the nearest trees or buildings.[6]

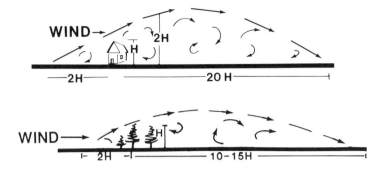

FIGURE 12.   Zones of disturbance and turbulence around buildings and windbreaks.

V. TYPES OF WIND MACHINES

Any device that converts the kinetic energy in the wind to a usable form of energy is called a wind machine. Many types of wind machines have been devised, and there are as many patents on wind machines as almost any other type of device. Basically, all wind machines remove kinetic energy from the wind by slowing it down and converting this energy to mechanical energy transmitted by a rotating shaft. Two basic types of machines have evolved and are classified as drag and lift types.

A. Drag Devices

These simple machines have flat, curved, or cup-shaped blades made of wood, steel, and/or other materials — the most common example being the American farm windmill. Another drag device is the cup anemometer used to measure windspeed. In both instances, the wind pushes on the blade or cup, forcing the rotor to turn about its axis (Figure 13A). This axis can be horizontal (parallel to the ground) as on the farm windmill or it can be vertical (perpendicular to the ground) as on the anemometer.

Drag devices characteristically produce high starting torque and are well suited to pumping water in low volumes. Experimenters have tried several approaches to improving their performance. One popular example is the "S" or Savonius rotor, often made from split oil drums. The farm windmill and similar drag devices are inherently limited in the amount of energy they can extract from the wind. At best, only a third of the energy available can be captured by such wind machines.[4]

B. Lift Devices

Lift devices use slender airfoils for blades rather than plates or cups. When the wind strikes one of these blades, it flows over and around the blade, creating lift. This lift pulls the blade about the axis of rotation, spinning the rotor much as lift supports an airplane wing in flight (Figure 13B).

In general, wind turbines that use lift are characterized by only a few (two, three, or four) blades in contrast to the multiple blades of drag devices. This at first seems mysterious, that a windmill with only two blades can operate more efficiently than one with a large number of blades. A modern wind turbine using airfoils can convert twice as much wind energy to useful work as a drag device of the same area. This is one of the reasons why the multibladed windmill has not been adapted to generate electricity.

1. Operating Characteristics

The amount of power produced by the wind turbine depends primarily on the windspeed and the intercepted area of the rotor. Most wind turbines do not begin producing usable

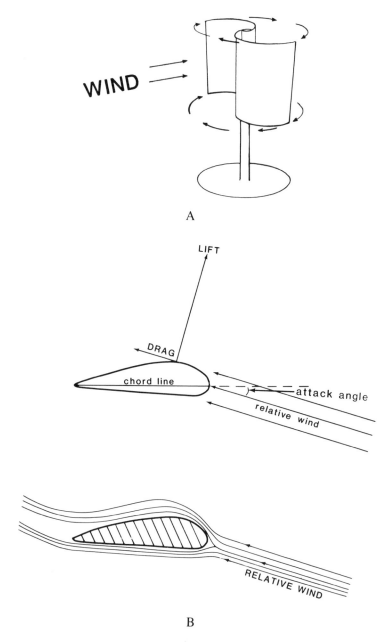

A

B

FIGURE 13.   (A) Drag-type wind machine. (B) Lift-type wind machine showing lift angle and drag force. (Reproduced with permission from *The Wind Power Book,* copyright © 1981 by Jack Park, published by Cheshire Books, Inc., Palo Alto, California.)

power until the windspeed reaches 4 to 5 m/sec. This beginning point is commonly referred to as the ''cut-in'' windspeed as shown in Figure 14. As windspeed increases above cut-in, power increases until rated power is reached. Rated power normally will occur between 10 and 15 m/sec, and is usually the peak output of the machine. If windspeed increases further, the power output of the generator is limited by a governing device that controls rotor speed. At some point, windspeeds are reached that can damage the wind machine and then the unit is automatically shut down. This high windspeed is called the ''cut-out'' windspeed. The

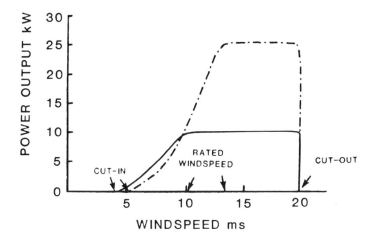

FIGURE 14. Typical power curves for wind turbines. The dashed curve represents a larger generator which has a higher cut-in windspeed and is less efficient at lower windspeeds. Rated windspeed is when the capacity of the generator is reached.

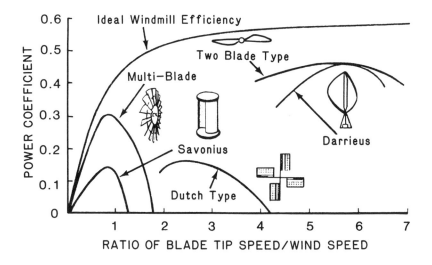

FIGURE 15. Typical performance curves for several types of wind machines. (Adapted from Eldridge, F. R., *Wind Machines,* 2nd ed., Van Nostrand Reinhold, New York, 1980.)

difference between cut-in and cut-out windspeeds becomes the operating range. Figure 14 represents a typical power curve for most wind machines, and the dashed line would represent a unit of larger diameter rather than one with more rotor blades.

Another term used to describe the performance of a wind machine is the tip-speed ratio. It is the ratio of the linear speed of the blade tip to the corresponding windspeed. Figure 15 shows some typical performance curves for different types of wind machines. The rotor efficiency or power coefficient, C_p, is the percent of available wind power that is extracted by the rotor. The drag devices (American multiblade and Savonius) have low tip-speed ratios, usually less than 1, while the lift devices have high tip-speed ratios.[11] Sometimes drag machines are also called low-speed machines and lift devices called high-speed machines because of this operating condition.

Four generic types of wind machines will be discussed to provide a better understanding of components and operating theory.

FIGURE 16. Water pumping windmills are used to provide water for livestock across much of the Southwest.

a. The Multiblade Farm Windmill

As stated in the beginning of this chapter, the multibladed farm windmill was developed in the mid-1800s and was designed primarily for pumping water from wells (Figure 16). The units manufactured today have changed little from the ones made in 1930 when an enclosed oil-lubricated gearbox was introduced. Figure 17 shows the typical components of a unit. Rotor diameter ranges from 2 to 6 m and units contain 16 to 18 curved-shaped metal blades. Air flow through the rotor causes a cascading effect, thus enhancing the drag and increasing rotor efficiency to 30%. The desirable features of this multibladed rotor are

- High starting torque
- Simple design and construction
- Simple control requirements
- Durability

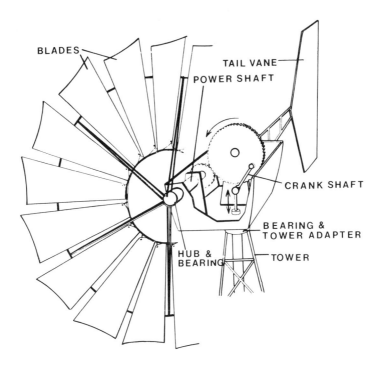

FIGURE 17. Essential components of a multibladed windmill. (Reproduced with permission from *The Wind Power Book,* copyright © 1981 by Jack Park, published by Cheshire Books, Inc., Palo Alto, California.)

The undesirable features are

- Exerts high rotor drag loads on tower
- Not readily adaptable to loads other than water pumping[4]

It is estimated that over 50,000 units are still operating in the southwestern U.S., and manufacturers have indicated that sales have been increasing since 1978. This is a good example of a complete system manufactured to perform a particular task.

b. The Savonius Rotor

The Savonius rotor, or S-rotor, was invented in the early 1920s and has received considerable attention because of its simple construction. Figure 18 shows the arrangement of the curved surfaces. The desirable features of the high drag S-rotor are

- Easily manufactured
- High starting torque

The undesirable features are

- Difficult to control because controls to limit rotational speed in high winds have not been devised
- Poor use of materials because it presents a small frontal area for fixed amount of construction materials[4]

This unit is best suited for pumping water, driving compressors or pond agitators, or other direct loads. Because of its slow rotational speed and control problems, it is not well suited to electrical generation.

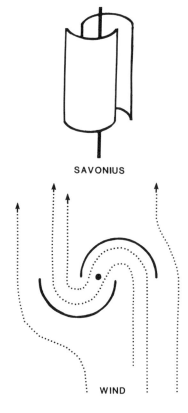

FIGURE 18.   The Savonius or S-rotor has several curved vanes to catch the wind.

c. *Horizontal Propeller Types*

These lift-type machines are the most common wind machine used for electrical generation and normally have two or three blades (Figure 19). Most machines are designed to produce a maximum efficiency of 40 to 45% at a tip-speed ratio of 5 to 6. Because of this higher operating speed, these units are almost always used for generation of electricity. The desirable features of these high-speed, propeller-type rotors are

- Slender blades use less material for same power output
- Higher rotational speeds reduce gearbox requirements
- Lower tower loads
- Larger diameter and high power levels are more easily obtained

The undesirable features are

- Low starting torque
- Blades require good aerodynamic design
- Possible flutter and vibration problems

An important feature of horizontal-axis machines is the manner in which they respond to changes in wind direction. Many small wind generators use a tail vane to keep the rotor perpendicular to the wind. Most large wind generators, and even some of the smaller ones, operate by placing their rotors downwind of the tower. The blades are angled slightly downwind (coned), making the turbine respond to changes in wind direction much like those

FIGURE 19. A modern horizontal-axis wind turbine used to produce electricity.

with tail vanes. This is why horizontal-axis wind turbines are also classified as upwind or downwind.

The major components of a horizontal-axis downwind machine are shown in Figure 20. Early manufacturers were expending a large percentage of their design efforts into perfecting the blades, but soon learned that bearings, gearboxes, and brakes require special attention too. The large number of operating hours per year and vibratory loads require specially designed components.

Electricity is transferred from the generator through slip-rings or twist cables to the ground. For the larger machines, twist cables have proven to be more reliable and cost effective. Because propeller machines have a low starting torque, they are usually not started under load, thus requiring an electronic control system to activate the load when the correct rotational speed is reached. In a similar manner, the control system must maintain a reasonable rotational speed in high winds. Therefore, the control system becomes an important part of the wind machine and often determines the success of a particular design.

d. Vertical-Axis or Darrieus

The Darrieus rotor was developed by the Frenchman G. J. M. Darrieus in the early 1920s. The Darrieus is classed as a vertical-axis unit, but differs from the Savonius in that it operates on aerodynamic lift principles. It has a tip-speed ratio of 6 to 7, and has a rotor efficiency of 40 to 45%. The desirable features of Darrieus or vertical-axis wind turbines are

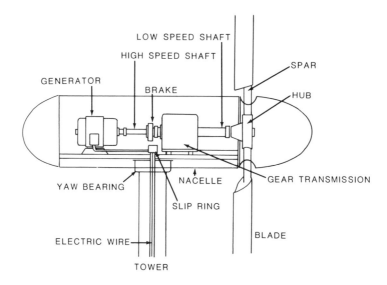

FIGURE 20.   Components of a typical horizontal-axis wind turbine with electric generator.

- Low materials usage for high power output
- High maintenance items are at ground level
- Can accept wind from any direction
- Mechanical power can be utilized at ground level

The undesirable features are

- Rotor is not normally self starting
- Requires larger land area to accommodate guy cables
- Needs good rotor control system to prevent rotor stall or overspeed

A Darrieus unit is shown in Figure 21 with its two symmetrical, airfoil-shaped blades. Recent data have indicated that large vertical-axis units may be more efficient and aerodynamically stable than horizontal-axis units because each blade increment on the vertical axis operates in the same wind shear region, whereas on a horizontal-axis unit, each blade increment rotates through several wind shear regions, causing extreme changes in blade loading per revolution. Unequal blade loading is a major problem on horizontal-axis machines in excess of 50 m diameter.

Vertical-axis machines normally have curved blades attached at each end to reduce the stresses in the blades. However, straight-bladed units have been built and tested, some with fixed blades and others with articulating blades that allow the machines to operate with variable pitch. Because of the large number of moving parts, these units have not been very reliable.

Wind machines come in many shapes and sizes, but all include the basic components of a rotor, power transmission, and control system. Regardless of the shape, power output is primarily determined by rotor area. A machine should be selected from a proven manufacturer (one that has sold at least 100 units), and should be large enough to provide the power required.

VI. APPLICATION IN AGRICULTURE

Wind power is ideally suited for many of the stationary uses that are now supplied by

FIGURE 21. Darrieus or vertical-axis wind turbine used to generate electricity.

utility electricity, natural gas, or liquified petroleum gas. Some examples of potential uses are water pumping, grain drying, milk cooling, water heating, and heating and ventilating of livestock buildings. Because most wind turbines produce either mechanical or electrical energy, the applications will be grouped into electrical and mechanical.

A. Electrical Loads

The most common use of wind energy is to generate electricity. Electricity is ideally suited to the varying amount of power in the wind because there is a simple load match between the wind rotor and the electric generator. A wind turbine can be located to maximize exposure to the wind with power lines to wherever the electricity is needed — your home, shop, barn, or pumps.

Most wind turbines being built or under development today are designed to generate electricity. They can be bought in almost any size, shape, and output — 120 to 440 V, DC or AC, varying or constant frequency. Some produce variable voltage and variable frequency while others provide 60-Hz AC power. In any case, the power from a wind generator can be changed or "conditioned" to meet the specific needs of the load.

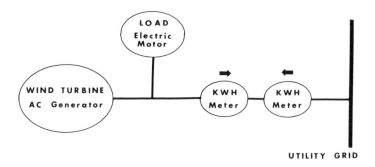

FIGURE 22.   Schematic of electrical connection for a wind-assisted pump using an induction generator.

1. Utility Intertie Systems

Wind systems that can operate in conjunction with the lines of the utility company have induction generators or alternators with synchronous inverters which convert DC to AC. In the operation of an induction generator, the field coils are excited by the utility line, and 60-Hz AC power is produced in phase with the utility line. The inverters take the electrical power from the wind generator and synchronize that power with the power from the utility (120 or 240 V, 60-Hz AC).[12]

When the wind generator produces less electricity than is required, the utility provides the difference, and when the wind generator produces more power than can be used, the excess is fed back into the utility lines (Figure 22). The power company, in effect, takes the place of storage batteries. It should be pointed out that the utility will normally pay less for the power they buy than they will charge for the electricity they sell.

The load on an electrical system containing a wind generating system is operated in a normal manner because the electrical loads cannot distinguish any difference between the utility-supplied electricity and the wind-generated electricity. However, there is an economic advantage to operate loads primarily when the wind is providing power because the amount of purchased power is reduced. When the utility line is off, no power is available from the wind turbine.

a. Induction Generator System

Most wind machines built for commercial sales in the U.S. are the induction-generator type. These units require that the generator be directly connected with the electric utility for excitation of the field coils. Most large induction generators are three-phase, 480-V systems.

Two wind systems with 25-kW induction generators have been operated to supply power for irrigation pumps at the USDA Conservation and Production Research Laboratory in Bushland, Tex. Both of these systems provided 480-V, three-phase power that was either used by the pumps or returned to the utility system. Table 2 summarizes the output data from the two systems. The Carter 25* was a 10-m diameter, two-bladed horizontal-axis machine (Figure 23) and the Enertech 44/25* was a 13.3-m diameter, three-bladed horizontal-axis machine (Figure 24). The Enertech had an effective area almost twice the size of the Carter; therefore, its output would be approximately double the Carter. Operating time is defined as the time the machines were on-line producing power, and little difference was observed between the two machines. The 64% operating time is considered good because the turbines do not start producing power until a 4.9 m/sec windspeed is reached.[13]

---

*   Trade names and manufacturer's model numbers are given for informational purposes only. Wind units used in USDA testing were purchased by USDA through competitive bids. No endorsement is given or implied to any manufacturer.

Table 2
SUMMARY OF PERFORMANCE DATA FOR CARTER 25 AND ENERTECH 44 WIND TURBINES (BUSHLAND, TEX.)

Month	Carter 25				Enertech 44				Windspeed @ 10 m
	Operating time		Energy produced	Availability	Operating time		Energy produced	Availability	
	hr	%	(kWhr)	(%)	hr	%	(kWhr)	(%)	(m/sec)
June 1982	405	50	3148	76	—	—	—	—	5.3
July	123	16	253	39	474	68	4947	100	5.6
August	423	59	1481	98	478	60	2984	100	4.6
September	503	72	2468	99	525	73	4536	100	5.5
October	418	67	2300	100	516	69	5023	98	5.5
November	373	44	2970	65	441	61	5292	98	6.3
December	255	36	2343	48	410	59	5135	99	6.8
January 1983	451	59	2574	100	286	37	2806	81	5.2
February	474	67	3606	100	344	49	3642	76	5.8
March	526	70	4224	100	512	69	5989	98	6.2
April	521	69	4280	95	476	64	5922	100	6.7
May	527	73	3883	100	537	68	6769	100	6.6
June	563	78	3710	100	539	75	5959	100	6.2
July	588	79	3483	99	584	79	6290	100	6.2
August	463	60	890	81	301	41	1836	82	4.4
September	543	80	3480	99	474	68	5823	81	6.4
October	583	78	3387	100	543	71	6753	100	5.8
November	513	69	3580	86	469	65	5979	92	6.3
Total	8251		52,060		7909		85,685		
Average	458	63	2892	88	465	64	5040	94	5.9

FIGURE 23. A 10-m diameter wind turbine with a 25-kW induction generator.

Availability is defined as the time that the control system of a machine is on and the machine is available to operate if sufficient winds are available. The 88% for the Carter and 94% for the Enertech are considered good for 1981 model machines. Both repairs and scheduled maintenance are included in the downtime. The delay in delivery of parts from the factory accounted for almost half of the downtime reported for these two units.

Clark[14] compared downtime in wind machines with other farm equipment and found that bearings, gearboxes, and other similar components failed at approximately the same rate. Bearing life was almost identical with that of electric motors and other similar equipment when based on number of revolutions rather than calendar days. The failure rate was more frequent in wind turbines because they operate almost three times as many hours per year as other equipment. Repairs and replacement of bearings should be performed every 18 months rather than 4 to 5 years as normally recommended.

Wind turbines operated with a primary load in conjunction with the electric utility are often referred to as wind-assist applications. An electrical wind-assist system using two wind turbines, with a combined rated output of 125 kW, was connected through a common load center with two irrigation pumps, a center-pivot sprinkler system, and an environmentally controlled data collection trailer. Energy production and use data for this load center are shown in Figure 25. The first bar for each month is a combination of wind-generated electricity and purchased electricity. Even though significantly more electricity was produced

FIGURE 24. A 13.3-m diameter wind turbine with a 25-kW induction generator.

than used, some electricity was purchased each month. Also, a significant percentage of wind-generated electricity was fed into the utility system, regardless of the amount used on site.

b. Synchronous Inverter

A few wind-electric generator systems use DC generators, self-excited alternators, or permanent magnet alternators. While these systems convert wind power to electricity differently, they each require a synchronous inverter to make the electrical output compatible with the electric utility. The synchronous inverter is normally a solid-state device that takes a variable DC voltage input and applies this energy to the AC line (Figure 26). Even if the wind turbine produced variable frequency, variable voltage AC power, it must be rectified before entering the inverter.

Synchronous inverters have the added advantage that they may operate with stored DC energy from batteries as long as the utility line voltage is present for synchronization. Electric power can be used from these systems for some applications (resistance heating) before entering the inverter because constant voltage and frequency are not required.

Operational experiences with synchronous inverters have shown that low line voltage due

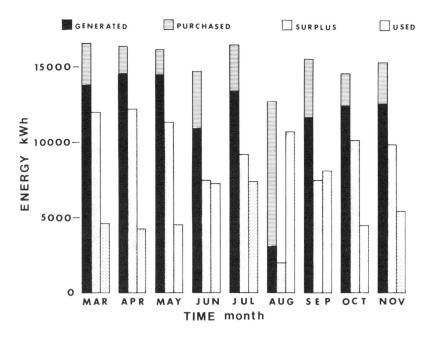

FIGURE 25.   Monthly energy generated by the Carter 25 and DOE 100, energy purchased from the utility, surplus sold to utility, and energy used by the irrigation system.

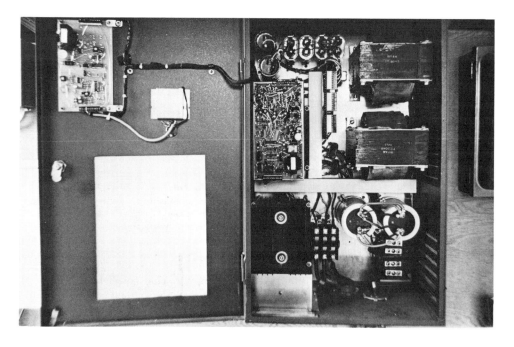

FIGURE 26.   A synchronous inverter used to convert DC electricity to AC power synchronous with the utility power.

to temporary power outages or lightning cause extensive damage to some solid-state components. Lightning suppressors have not corrected the problem. Performance data indicate that the inverter has good efficiency at its rated power, but drops off sharply when operating below 50% of rated output. This is significant in wind power application because wind turbines rarely operate near rated power. Other disadvantages of a synchronous inverter are the introduction of harmonics (electrical resonance peaks) which cause electrical interference and high noise levels if placed near a workspace.[15]

2. Nonintertied Electrical Systems

Although a majority of modern wind turbines are designed for interconnection to the utility grid, there are a number of agricultural applications that do not require utility-grade power.

Providing heat for 15.5 million rural residences (houses outside corporate limits) requires over one fourth of the energy used on farms. Heating oil, liquid petroleum gas, and natural gas supply fuel for 80% of these rural residences. An autonomous (nonintertied) wind turbine, providing power for resistance heaters, could provide as much as 75% of the heating requirement when a small heat storage system is incorporated with the wind turbine system. Electrical resistance heaters do not require 60-Hz AC power; therefore, they are suitable for use with variable voltage and frequency. The wind-generated electricity is supplied to the resistance heaters, which in turn are used to heat water. The hot water is used for heating the air space or as a preheating for the hot water supply. The major advantage of this heating system is that it continues to operate when power failures occur.[16]

USDA has also conducted studies with autonomous wind systems to heat water for dairies and grain drying. In both cases, the heating system served as a preheating or initial heater. For dairy water heating, water was rarely heated above 50°C, but made a significant reduction in total energy used. For grain drying, the concept of low-temperature drying worked well with wind-powered heat sources. Almost all heating concepts utilize resistance heaters operating at variable voltage and frequency.[16]

Water pumping for low-lifts has been accomplished by using a permanent magnet alternator wind energy conversion system designed to operate with a rotor speed from 70 to 150 rpm. The wind turbine with variable-voltage, variable-frequency electrical output was used to power an autonomous pumping system. The frequency of the output varied from 30 to 65 Hz while the voltage changed from 85 to 250 V, resulting in a voltage to frequency ratio from 2.6 to 3.3. Laboratory tests revealed that standard 240-V, three-phase, AC induction motors would pump water with the variable speed system if the voltage to frequency ratio was fairly constant. Figure 27 shows the pump curves for a centrifugal pump when powered by the utility and the permanent magnet alternator. Overall efficiency of the variable speed system approached that of a conventional system.[17]

Other possible uses of autonomous electric wind generators are for converting sea and brackish water to potable water and meeting some heating or cooling loads in the food processing industry. Because of the seasonal use of many of these applications, the economics of installations for a single purpose do not appear desirable.

B. Mechanical

The most common type of mechanical wind turbine is the American multibladed water pumping windmill described earlier. As with all wind systems the generator, or pump in this case, is an integral part of the system. If any part of the system is significantly modified, the output is usually decreased, resulting in unsatisfactory performance.

The pumps used with the multiblade windmill are piston-type with a plunger moving up and down within a fixed cylinder (Figure 28). The pumping capacity of common pump sizes is shown in Table 3 for different windmill sizes. This table is based on machines operating

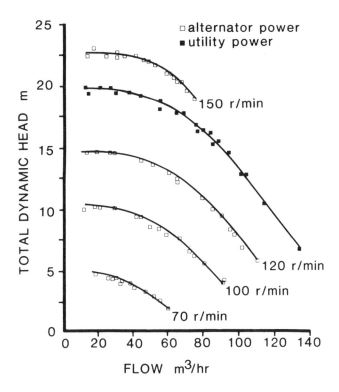

FIGURE 27.   Pump discharge rate, total dynamic head relationship for a centrifugal pump with a 7.6-kW electric motor supplied from utility and autonomous wind turbine.

at a 7 to 8 m/sec windspeed and using the long stroke length. If the short stroke is selected, the flow should be reduced by 25%. As can be seen by data in Table 3, the maximum flow rate of these systems is limited primarily by the pump diameter and stroke rather than by pumping lift or windspeed. Larger diameter rotors are selected only when more lift is needed.[18]

Since the multiblade water pumping system is limited to a small flow rate, it is used primarily for domestic and livestock water. The typical system with a 2.44-m rotor and 4.76-cm cylinder will supply sufficient water for a family of four and 15 to 20 large animals. When used for irrigation, the area is usually limited to a vegetable garden.

Irrigation pumping requires large amounts of power because crops like corn, rice, cotton, and wheat transpire 1 cm/day, thus requiring a flow of 1 ℓ/sec/ha. This is the amount of water that must be available throughout the growing season. Irrigators prefer to have between 30 and 50 ℓ/sec available from their pumps. Power requirements range between 10 and 150 kW, depending on the lift and discharge pressure. In 1980, farms in the U.S. used an estimated 90 billion kWh of energy for irrigation pumping. Electricity, natural gas, and diesel fuel were the major forms of energy used. Irrigation pumping energy accounts for 40 to 70% of the energy used on farms where irrigation is practiced.

Wind-powered irrigation systems have been developed by the USDA Conservation and Production Research Laboratory at Bushland, Tex. to supply at least 15 ℓ/sec to irrigated crops. These systems include ones that incorporate conventional power sources with wind power, called wind-assist, and stand-alone wind-powered pumps.

1. Mechanical Wind-Assist Pumping

This type of pumping system used both a vertical-axis wind turbine and a diesel engine

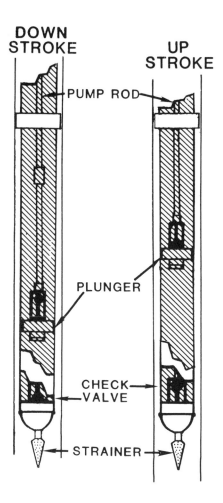

FIGURE 28. Schematic of a piston pump showing operation of the plunger.

Table 3
PUMPING CAPACITIES AND LIFT ELEVATIONS FOR MULTIBLADED WINDMILLS

| Cylinder size (cm) | Pumping capacity (m^3/hr) | Rotor diameter (m) | | | | |
		2.44	3.05	3.66	4.27	4.88
		Elevation (m)				
4.8	0.68	53	79	119	171	280
5.7	0.98	34	52	76	110	180
7.0	1.45	24	37	55	79	129
9.5	2.76	—	20	30	44	70

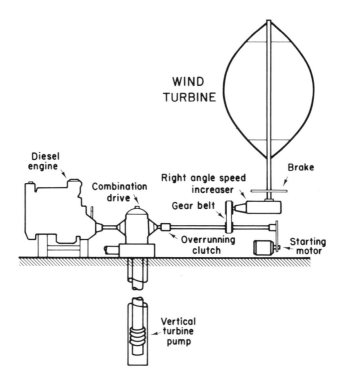

FIGURE 29. Schematic of a mechanical drive, wind-assist irrigation pumping system.

to power a vertical turbine pump (Figure 29). The diesel engine was sized to operate the pump on a stand-alone basis and ran continuously when irrigating. The wind turbine was coupled to the pump shaft through an overrunning clutch and combination gear drive, and furnished power only when the windspeed exceeded 5 m/sec. The wind turbine thus reduced the load on the diesel engine rather than replacing it.

The verical-axis wind turbine was 17 m high and had an equatorial diameter of 11.4 m and was rated at 40 kW. The turbine operated at 81 rpm and utilized a speed-increasing gearbox and timing belt to raise the shaft speed to 1780 rpm. The high-speed shaft was connected to a combination gear drive through an overrunning clutch which was used to synchronize the speeds of the wind turbine and the diesel engine. The pump delivered 20 ℓ/sec against a 106-m head and required 45 kW. When the windspeed was below 5 m/sec, the diesel engine supplied all the power to the pump and consumed about 3.8 mℓ/sec of fuel.

The system performance is summarized in Figure 30 for windspeeds ranging from 5 to 20 m/sec. The turbine power and diesel power were measured with separate torque transducers in the high-speed drive shafts. Notice that as the windspeed increased and the wind turbine produced more power, the load on the diesel decreased. This reduced load on the diesel engine resulted in reduced fuel consumption. At an average windspeed of 12 m/sec, the fuel consumption was reduced to 2.3 mℓ/sec or a 40% reduction. The fuel savings was estimated at 30% when pumping was done from March through October.[19]

This system has worked well with few operational problems. Two major disadvantages of this system are that the equipment can be used only to pump water and the system is difficult to automate.

2. Mechanical Stand-Alone Pumping

A small, vertical-axis wind turbine has been mechanically coupled to a positive displace-

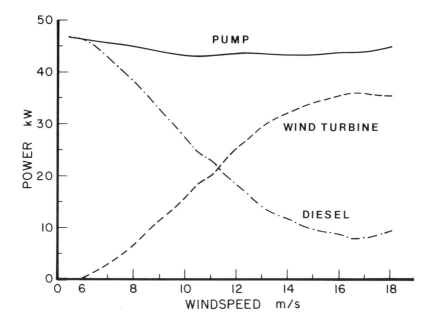

FIGURE 30. Mechanical wind-assist pumping system power, rotor power, and diesel engine power for a pumping rate of 20 ℓ/sec against a lift of 106 m.

ment pump for pumping from shallow wells or lakes. The 4-kW wind turbine was 5.5 m high and had an equatorial diameter of 4.5 m. Power was transferred through an electric clutch, which allowed the turbine to be started in a no-load condition (Figure 31). The rotor operated between 160 and 240 rpm, with corresponding flows of 5 to 12 ℓ/sec at a lift of 30 m. When the rotor reached a speed of 250 rpm, aerodynamic brakes deployed, slowing the rotor and preventing a runaway.

The rotational speed of the pumping system is shown in Figure 32 as a function of windspeed. The pump provided sufficient load so that rotational speeds increased slightly with increased winds. The stalling of the airfoil at windspeeds below 11 m/sec is shown by these data, indicating that the pump overloaded the turbine at low windspeeds. The need for an adjustable pumping load to match the windspeed is shown by these data because a small load is needed in low to moderate winds and a large load is needed in gusty high winds.[20]

3. Mechanical Water Heating

A wind-driven direct water heating system was developed for heating water for dairies. A mechanical-drive, vertical-axis wind turbine was connected to a paddle-like agitator immersed in water. As the agitator violently stirred the water, friction caused heat to be produced in the water. Since a wind turbine and a fluid agitator are both characteristically fan-in-fluid processes, they operate with a cubic power/speed relationship. This allows the agitator to fully utilize all of the mechanical output of the wind turbine at any rotational speed. An average 20°C temperature rise was common with this mechanical water heating system.[21]

VII. ECONOMICS

The value of any piece of agricultural equipment is difficult to determine because of the many and varied uses of that equipment. Wind machines have a different value to the farmer who is several miles from the nearest utility compared with the farmer where utility power is readily available. Economic comparisons are best made considering wind machines that

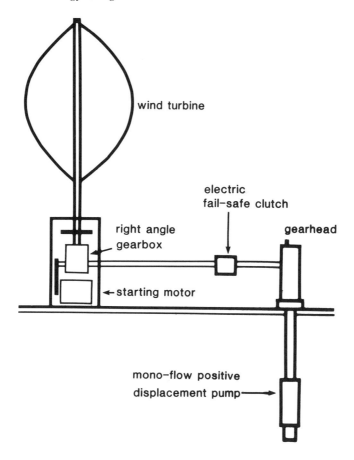

FIGURE 31. Schematic of a 4-kW stand-alone mechanical pumping system.

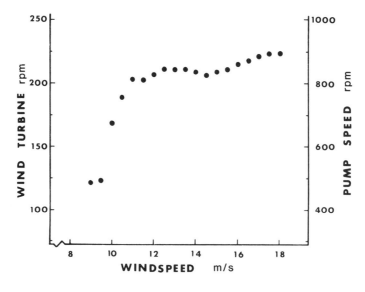

FIGURE 32. Wind turbine rotational speed and pump speeds as a function of windspeed for a 4-kW stand-alone pumping system.

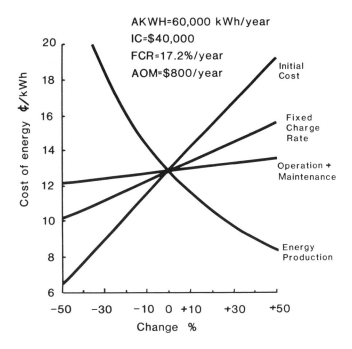

FIGURE 33.   Sensitivity diagram for cost of energy using data from an
Enertech 44/25 wind turbine.

are intertied with the utility. The cost of electricity produced by the wind turbine is easily
compared with the utility cost. When considering the utility electric cost, the demand charges
must be added to the energy cost to determine the total cost of electrical energy.

The cost of electrical energy produced by a wind turbine can be calculated from the
following equation:[22]

$$COE = [(IC)\ (FCR)\ +\ (AOM)]/AKWH \qquad (5)$$

where COE is the cost of energy (cents per kilowatt-hour), IC is the initial installed cost
(dollars), FCR is the fixed charge rate (percent per year), AOM is the annual operation and
maintenance (percent of IC per year), and AKWH is the net annual kilowatt-hour energy
production (kilowatt-hours per year).

As an example, if the initial cost of a system was $36,500, with an additional $3,500 for
installation and a payback period of 10 years was selected with an annual rate of interest of
12%, this would result in a fixed charged rate of 17.2. Annual operation and maintenance
was assumed to be 2% of the initial cost, which is $800. An annual net energy production
of 60,000 kWh was used for the single-phase generator. The cost of energy with the preceding
data was calculated to be 12.8 ¢/kWh.

The actual initial cost of wind machines varies considerably from state to state because
of different tax credits and incentives for development of alternative energy sources. Federal
tax credits are available through 1985 for individuals and businesses who purchase and install
wind machines. State credits vary from 0 to 30%; therefore, the same machine may actually
cost one owner less than a nearby owner because of state differences. In order to easily
determine the effect of tax changes and the importance of the other parameters that affect
the cost of energy, sensitivity curves were developed for the example above (Figure 33).
From these curves, it is easy to determine how each parameter influences the cost of energy.
The critical factors are initial cost and annual energy production. A 10% decrease in initial

cost or a 10% increase in energy production would result in a 2 ¢/kWh drop in the cost of energy. For this example, interest and maintenance have the least effect and therefore should not be considered as much as annual energy production and initial cost. Energy production is primarily a function of the site (windspeed, air density, and duration of wind) and the availability of the machine.[23]

VIII. SAFETY

Wind energy systems are much like other pieces of machinery in that they can be extremely dangerous if not used carefully. Any safety precautions used with other agricultural machinery would be applicable to wind turbines. Hazards to personnel generally occur during erection, operation, or servicing, while hazards to equipment normally occur during storms.

All wind turbines are placed on towers to raise the rotors into the clear wind stream, thus avoiding turbulence from trees and buildings. Tower heights vary from 10 m for water-pumping windmills to 40 m for some larger electric generators. Presently, most new units are being installed on 25- to 30-m towers. Vertical-axis units normally have smaller support stands (3 to 10 m).

A. Personnel Safety

Steel workers and electric power linemen all use work belts or harnesses when working on towers or large steel structures. The belt serves as a safety device to prevent the worker from falling. The "sit harness," developed for sport climbing, is becoming the preferred aid when working on towers. Harnesses are preferable to belts because they are easy to use and are more comfortable when hanging from a tower for long periods.[24] In addition to belts or harnesses, lanyards and snap hooks are needed to complete your climbing equipment. These items are needed to secure the worker and equipment in place while working. It is important that the rotor, nacelle, etc. not be moving while servicing because of the possibility of striking a worker. Also, many accidents occur while workers are ascending or descending the tower. A climbing safety device is essential to keep the climber attached to the tower at all times and prevent his falling.

Boots and gloves also aid the climber by giving him a better grip and reducing fatigue. Boots reduce the fatigue on feet and legs while gloves help to improve the grip. Hard hats are helpful to the worker when dealing with machines without parking brakes.

Some good rules to remember are (1) never climb the tower while the turbine is running, (2) always apply the brake, furl the tail, or do whatever is needed to stop the unit, (3) never work alone, and (4) use your safety equipment — it does no good if you do not wear it.

B. Equipment

Since about 95% of all wind machines sold today are used to produce electricity, the hazards of electrical shock are present. At the generator, the power is converted from rotating mechanical energy to electrical energy; thus, hazards from both rotating shafts and electricity exist. You must be careful to watch for both hazards.

Induction generators require a connection with the utility grid for field excitation; therefore, the generator is at the same voltage as the utility grid. A manual disconnect is necessary to separate the generator from the utility during maintenance and service. This not only protects the person working on the wind turbine, but also protects the utility serviceman when working on the utility lines. The synchronous inverter also requires hook-up with the utility grid; therefore, it also needs a disconnect. You must observe the same precautions and safety practices as you would when working on any electrical device. All wiring and equipment should adhere to the National Electric Code and local requirements. Service personnel must make sure the power is off before servicing the unit.

Tall, well-grounded metal objects attract lightning, and wind turbines are no exception. Lightning causes more deaths each year than tornadoes and offers the potential for accidents with wind turbines. At the present time, no one has been killed or seriously injured near a wind turbine due to lightning; however, personnel should not work on towers or be near them during storms containing lightning.

Most damage attributed to lightning has been caused to the wind turbine control systems. Direct lightning strikes or nearby strikes, carried along electrical lines, cause severe problems to solid-state control devices and other electrical controls. The use of lightning arrestors often helps to minimize the damage, but most arrestors are not sufficient to totally protect the system. It is recommended that a lightning arrestor be placed at each meter, disconnect, and controller. If a microprocessor (small computer) is used in the control system, then its power supply should contain a current and voltage surge protector. The use of lightning arrestors can reduce damage by over 50%.

All wind turbines should be adequately grounded by following the National Electric Code or equivalent guidelines. Several ground rods should be used and intertied to prevent a temporary loss of ground, as often occurs in dry areas with only one rod. Controllers often detect ground loops when all parts of the system are not connected to a common ground. Ground loops can cause erroneous signals to be transmitted to the controller, thus confusing the controller.

REFERENCES

1. **Golding, E. W.,** *The Generation of Electricity by Wind Power,* E. & F. N. Spon, London, 1955, 332.
2. **Gates, B.,** Blowing in the wind, *Tex. Highways,* 28(3), 4, 1981.
3. **Nelson, V.,** A history of the SWECS industry in the U.S., *Alternative Sources of Energy,* 66, 20, 1984.
4. **Park, J.,** *The Wind Power Book,* Cheshire Books, Palo Alto, Calif., 1981, 253.
5. **Clark, R. N.,** Data Requirements for Wind Energy, ASAE Paper No. 80-4517, American Society of Agricultural Engineers, St. Joseph, Mich., 1980.
6. **Wegley, H. L., Ramsdell, J. V., Orgill, M. M., and Drake, R. L.,** A Siting Handbook for Small Wind Energy Conversion Systems, Report No. PNL-2521 Rev1, U.S. Department of Energy, Washington, D.C., March 1980.
7. Battelle Pacific Northwest Laboratory, United States Annual Average Wind Power, U.S. Department of Energy, Washington, D.C., 1981.
8. **Elliott, D. L. and Barchet, W. R.,** *Wind Energy Resource Atlas:* Vols. 1 to 12, Battelle Pacific Northwest Laboratory, 1981.
9. **Clark, R. N.,** Wind conversion equipment and utility interties, in *Proc. Solar and Wind Systems Workshop,* Great Plains Agricultural Council Publication No. 108, Lincoln, Neb., 1983.
10. **Hewson, E. W., Wade, J. E., and Baker, R. W.,** Vegetation as an Indicator of High Wind Velocity, Report No. RLO-2227, U.S. Department of Energy, Washington, D.C., June 1979.
11. **Eldridge, F. R.,** *Wind Machines,* 2nd ed., Van Nostrand Reinhold, New York, 1980, 214.
12. **Clark, R. N.,** Generation of Electricity with Wind Power, ASAE Paper No. 80-3521, American Society of Agricultural Engineers, St. Joseph, Mich., 1980.
13. **Clark, R. N. and Vosper, F. C.,** Electrical wind assist water pumping, in *Proc. 1984 ASME Wind Energy Symposium,* New Orleans, La., 1984.
14. **Clark, R. N.,** Reliability of Wind Electric Generators, ASAE Paper No. 83-3505, American Society of Agricultural Engineers, St. Joseph, Mich., 1983.
15. **Soderholm, L. H. and Clark, R. N.,** Wind Energy Concepts as Related to Parallel Generation, ASAE Paper No. 81-3003, American Society of Agricultural Engineers, St. Joseph, Mich., 1981.
16. **Soderholm, L. H.,** Non-interconnected application of wind energy, in *Proc. Rural Electric Wind Energy Workshop,* REA, Boulder, Colo., 1982, 285.
17. **Vosper, F. C. and Clark, R. N.,** Electrical stand-alone water pumping, in *Proc. of Wind Workshop VI,* Minneapolis, Minn., 1983, 741.
18. New Mexico Energy Institute, Selecting Water-Pumping Windmills, New Mexico State University, Las Cruces, 1978, 14.

19. **Clark, R. N.,** Co-generation using Wind and Diesel for Irrigation Pumping, ASAE Paper No. 84-2603, American Society of Agricultural Engineers, St. Joseph, Mich., 1984.

20. **Clark, R. N.,** Irrigation pumping with wind energy only, *Agric. Eng.,* 64(12), 15, 1983.

21. **Gunkel, W. W., Lacey, D. R., Neyeloff, S., and Porter, T. G.,** Wind Energy for Direct Water Heating, Report No. DOE/SEA 3408-20691-81-2, U.S. Department of Energy, Washington, D.C., 1981.

22. **Ramler, J. R. and Donovan, R. M.,** Wind Turbines for Electric Utilities: Development Status and Economics, NASA Report No. TM-79170, 1979.

23. **Vosper, F. C. and Clark, R. N.,** Operation of a Third Generation Wind Turbine, ASAE Paper No. 83-4542, American Society of Agricultural Engineers, St. Joseph, Mich., 1983.

24. **Gipe, P. and Schillmoeller, S.,** Working with the wind: tower safety, *Alternative Sources of Energy,* May/June, 1983.

Chapter 4

BIOMASS GASIFICATION

A. K. Rajvanshi

TABLE OF CONTENTS

I. INTRODUCTION

Modern agriculture is an extremely energy-intensive process. However, high agricultural productivities and the subsequent growth of the green revolution have been made possible only by large amounts of energy inputs, especially those from fossil fuels.[1] With recent price rises and the scarcity of these fuels there has been a trend towards the use of alternative energy sources like solar wind, geothermal, etc.[2] However, these energy resources have not been able to provide an economically viable solution for agricultural applications.[3]

One biomass energy-based system which has been proven reliable and had been extensively used for transportation and on farm systems during World War II is wood or biomass gasification.[4]

Biomass gasification means incomplete combustion of biomass resulting in production of combustible gases consisting of carbon monoxide (CO), hydrogen (H_2), and traces of methane (CH_4). This mixture is called *producer gas*. Producer gas can be used to run internal combustion engines[4] (both compression and spark ignition), can be used as a substitute for furnace oil in direct heat applications, and can be used to produce, in an economically viable way, methanol — an extremely attractive chemical which is useful both as fuel for heat engines as well as chemical feedstock for industries.[5] Since any biomass material can undergo gasification, this process is much more attractive than ethanol production or biogas where only selected biomass materials can produce the fuel.

Besides, there is a problem that solid wastes (available on the farm) are seldom in a form that can be readily utilized economically, e.g., wood wastes can be used in a hog fuel boiler but the equipment is expensive and energy recovery is low.[6] As a result, it is often advantageous to convert this waste into a more readily usable fuel form like producer gas. Hence the attractiveness of gasification.

However, under present conditions, economic factors seem to provide the strongest argument for considering gasification.[7,8] In many situations where the price of petroleum fuels is high or where supplies are unreliable, the biomass gasification can provide an economically viable system — provided the suitable biomass feedstock is easily available (as is indeed the case in agricultural systems).

II. HISTORICAL BACKGROUND

The process of gasification to produce combustibles from organic feeds was used in blast furnaces over 180 years ago. The possibility of using this gas for heating and power generation was soon realized and there emerged in Europe producer gas systems which used charcoal and peat as feed material. At the turn of the century, petroleum gained wider use as a fuel, but during both World Wars, and particularly World War II, a shortage in petroleum supplies led to widespread reintroduction of gasification. By 1945 the gas was being used to power trucks, buses, and agricultural and industrial machines. It is estimated that there were close to 900,000 vehicles running on producer gas world-wide.[9]

After World War II the lack of strategic impetus and the availability of cheap fossil fuels led to a general decline in the producer gas industry. However, Sweden continued to work on producer gas technology and the work was accelerated after the 1956 Suez Canal crisis. A decision was then made to include gasifiers in Swedish strategic emergency plans. Research into suitable designs of wood gasifiers, essentially for transport use, was carried out at the National Swedish Institute for Agricultural Machinery Testing and is still in progress.[10]

The contemporary interest in small-scale gasifier research and development for most part dates from the 1973 oil crisis. The U.S. research in this area is reviewed by Goss.[11] Manufacturing also took off, with increased interest shown in gasification technology. At present there are about 64 gasification equipment manufacturers world-wide.[11,36] The present

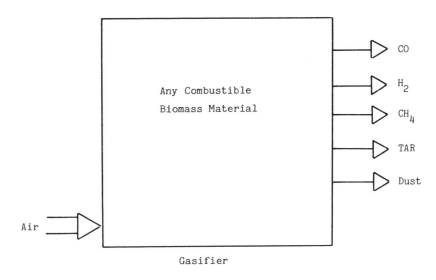

FIGURE 1.   Products of gasification.

status of gasification technology and research and development activities will be discussed in Section VII.

III. THEORY OF GASIFICATION

The production of generator gas (producer gas) called *gasification,* is the partial combustion of solid fuel (biomass) and takes place at temperatures of about 1000°C. The reactor is called a *gasifier.*

The combustion products from complete combustion of biomass generally contain nitrogen, water vapor, carbon dioxide, and a surplus of oxygen. However, in gasification, where there is a surplus of solid fuel (incomplete combustion), the products of combustion are (Figure 1) combustible gases like carbon monoxide (CO), hydrogen (H_2), and traces of methane and nonuseful products like tar and dust. The production of these gases is by reaction of water vapor and carbon dioxide through a glowing layer of charcoal. Thus, the key to gasifier design is to create conditions such that biomass is reduced to charcoal and charcoal is converted at suitable temperature to produce CO and H_2.

A. Types of Gasifiers

Since there is an interaction of air or oxygen and biomass in the gasifier, they are classified according to the way air or oxygen is introduced in it. There are three types of gasifiers (Figure 2): downdraft, updraft, and crossdraft. As the classification implies, an updraft gasifier has air passing through the biomass from the bottom and the combustible gases come out from the top of the gasifier. Similarly, in the downdraft gasifier the air is passed from the tuyers in the downdraft direction. With slight variation almost all the gasifiers fall in the above three categories.

The choice of one type of gasifier over other is dictated by the fuel, its final available form, its size, moisture content, and ash content.[12] Table 1 lists the advantages and disadvantages generally found for various classes of gasifiers.[13]

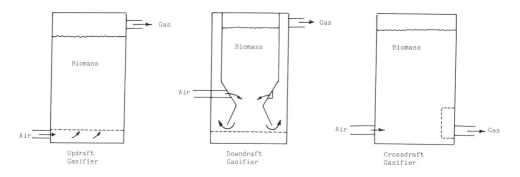

FIGURE 2.   Various types of gasifiers.

Table 1
ADVANTAGES AND DISADVANTAGES OF
VARIOUS GASIFIERS

Gasifier type	Advantages	Disadvantages
Updraft	Small pressure drop Good thermal efficiency Little tendency towards slag formation	Great sensitivity to tar and moisture content of fuel Relatively long time required for startup of internal combustion engine Poor reaction capability with heavy gas load
Downdraft	Flexible adaptation of gas production to load Low sensitivity to charcoal dust and tar content of fuel	Design tends to be tall Not feasible for very small particle size of fuel
Crossdraft	Short design height Very fast response time to load Flexible gas production	Very high sensitivity to slag formation High pressure drop

B. Process Zones

Four distinct processes take place in a gasifier as the fuel makes its way to gasification:
1. Drying of fuel
2. Pyrolysis — a process in which tar and other volatiles are driven off
3. Combustion
4. Reduction

Though there is a considerable overlap of the processes, each can be assumed to occupy a separate zone where fundamentally different chemical and thermal reactions take place. Figure 3 shows schematically an updraft gasifier with different zones and their respective temperatures. Figures 4 and 5 show these regions for downdraft and crossdraft gasifiers, respectively.

In the downdraft gasifiers there are two types: single throat and double throat (Figure 6). Single throat gasifiers are mainly used for stationary applications whereas double throat models are for varying loads as well as automotive purposes.

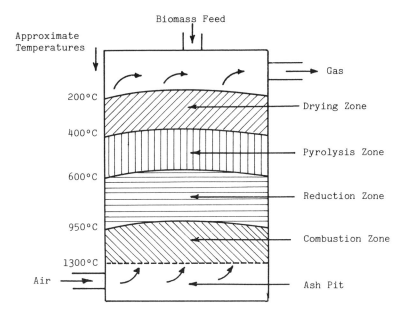

FIGURE 3. Various zones in updraft gasifier. (From Foley, G. and Barnard, G., Biomass Gasification in Developing Countries, Technical Report No. 1, Earthscan, London, 1983. With permission.)

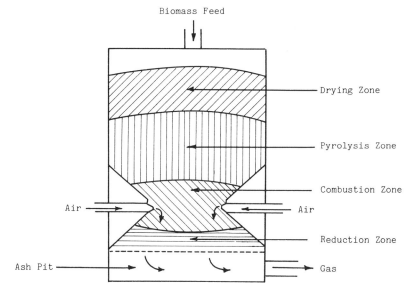

FIGURE 4. Gasification process in downdraft gasifier. (From Foley, G. and Barnard, G., Biomass Gasification in Developing Countries, Technical Report No. 1, Earthscan, London, 1983. With permission.)

C. Reaction Chemistry

The following major reactions take place in combustion and reduction zones.[12]

1. Combustion Zone

The combustible substance of a solid fuel is usually composed of the elements carbon, hydrogen, and oxygen. In complete combustion carbon dioxide is obtained from carbon in fuel and water is obtained from the hydrogen, usually as steam. The combustion reaction

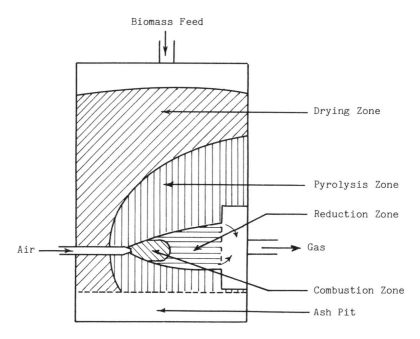

FIGURE 5.   Gasification process in crossdraft gasifier. (From Foley, G. and Barnard, G., Biomass Gasification in Developing Countries, Technical Report No. 1, Earthscan, London, 1983. With permission.)

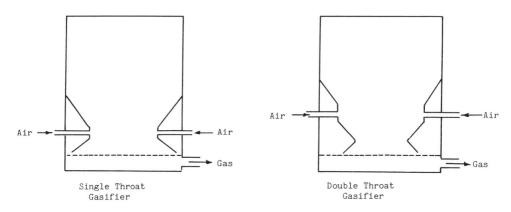

FIGURE 6.   Single and double throat gasifiers.

is exothermic and yields a theoretical oxidation temperature of 1450°C.[14] The main reactions, therefore, are

$$C + O_2 \quad = CO_2 \qquad (+ 393 \text{ MJ/kg mole}) \tag{1}$$

$$2H_2 + O_2 = 2H_2O \qquad (- 242 \text{ MJ/kg mole}) \tag{2}$$

2. Reduction Zone

The products of partial combustion (water, carbon dioxide, and uncombusted partially cracked pyrolysis products) now pass through a red-hot charcoal bed where the following reduction reactions take place:[12]

$$C + CO_2 \quad = 2CO \qquad (- 164.9 \text{ MJ/kg mole}) \tag{3}$$

$$C + H_2O \quad = CO + H_2 \qquad (- 122.6 \text{ MJ/kg mole}) \tag{4}$$

$$CO + H_2O = CO_2 + H_2 \qquad (+ 42 \text{ MJ/kg mole}) \tag{5}$$

$$C + 2H_2 = CH_4 \qquad (+ 75 \text{ MJ/kg mole}) \qquad (6)$$

$$CO_2 + H_2 = CO + H_2O \qquad (- 42.3 \text{ MJ/kg mole}) \qquad (7)$$

Reactions 3 and 4 are main reduction reactions and, being endothermic, have the capability of reducing gas temperature. Consequently, the temperatures in the reduction zone are normally between 600 and 700°C.

3. Pyrolysis Zone

Wood pyrolysis is an intricate process that is still not completely understood.[14] The products depend upon temperature, pressure, residence time, and heat losses. However, the following general remarks can be made about them.

Up to a temperature of 200°C only water is driven off. Between 200 to 280°C carbon dioxide, acetic acid, and water are given off. The real pyrolysis, which takes place between 280 to 500°C, produces large quantities of tar and gases containing carbon dioxide. Besides light tars, some methyl alcohol is also formed. Between 500 to 700°C the gas production is small and contains hydrogen.

Thus it is easy to see that an updraft gasifier will produce much more tar than a downdraft one. In a downdraft gasifier the tars have to go through the combustion and reduction zones and are partially broken down.

Since the majority of fuels like wood and biomass residue do have large quantities of tar, the downdraft gasifier is preferred over others. Indeed, the majority of gasifiers, both in World War II and presently, are of the downdraft type.

Finally, in the drying zone the main process is the drying of wood. Wood entering the gasifier has a moisture content of 10 to 30%. Various experiments on different gasifiers in different conditions have shown that, on the average, the condensate formed is 6 to 10% of the weight of gasified wood.[14] Some organic acids also come out during the drying process. These acids give rise to corrosion of gasifiers.

D. Properties of Producer Gas

Producer gas is affected by various processes as outlined above; hence, one can expect variations in the gas produced from various biomass sources. Table 2 lists the composition of gas produced from various sources. The gas composition is also a function of gasifier design and thus, the same fuel may give different calorific values when used in two different gasifiers. Table 2, therefore, shows approximate values of gas from different fuels.

The maximum dilution of the gas takes place because of the presence of nitrogen. Almost 50 to 60% of the gas is composed of noncombustible nitrogen. Thus it may be beneficial to use oxygen instead of air for gasification. However, the cost and availability of oxygen may be a limiting factor in this regard. Nevertheless, where the end product is methanol — a high-energy quality product — the cost and use of oxygen can be justified.[5]

On an average, 1 kg of biomass produces about 2.5 m³ of producer gas at S.T.P. In this process it consumes about 1.5 m³ of air for combustion.[14] For complete combustion of wood, about 4.5 m³ of air is required. Thus, biomass gasification consumes about 33% of the theoretical stoichiometric ratio for wood burning.

The average energy conversion efficiency of wood gasifiers is about 60 to 70% and is defined as:

$$\eta_{Gas} = \frac{\text{Calorific value of gas/kg of fuel}}{\text{Avg calorific value of 1 kg of fuel}} \qquad (8)$$

Table 2
COMPOSITION OF PRODUCER GAS FROM VARIOUS FUELS

| Fuel | Gasification method | Volume percentage | | | | | Calorific value (MJ/m³) | Ref. |
		CO	H₂	CH₄	CO₂	N₂		
Charcoal	Downdraft	28—31	5—10	1—2	1—2	55—60	4.60—5.65	12
Wood with 12 to 20% moisture content	Downdraft	17—22	16—20	2—3	10—15	45—50	5.00—5.86	12
Wheat straw pellets	Downdraft	14—17	17—19	—	11—14	—	4.50	15
Coconut husks	Downdraft	16—20	17—19.5	—	10—15	—	5.80	15
Coconut shells	Downdraft	19—24	10—15	—	11—15	—	7.20	15
Pressed sugarcane	Downdraft	15—18	15—18	—	12—14	—	5.30	15
Charcoal	Updraft	30	19.7	—	3.6	46	5.98	16
Corn cobs	Downdraft	18.6	16.5	6.4	—	—	6.29	17
Rice hulls, pelleted	Downdraft	16.1	9.6	0.95	—	—	3.25	17
Cotton stalks, cubed	Downdraft	15.7	11.7	3.4	—	—	4.32	17

For example, 1 kg of wood produces 2.5 m³ of gas with an average calorific value of 5.4 MJ/m³. The average calorific value of wood (dry) is 19.8 MJ/kg.[18] Hence,

$$\eta_{Gas} = \frac{2.5 \ (m^3) \times 5.4 \ (MJ/m^3)}{19.80 \ (MJ/kg) \times 1 \ (kg)} = 68\%$$

E. Temperature of Gas
On an average, the temperature of gas leaving the gasifier is about 300 to 400°C.[16] If the temperature is higher than this (~ 500°C) it is an indication that partial combustion of gas is taking place. This generally happens when the air flow rate through the gasifier is higher than the design value.

IV. GASIFIER FUEL CHARACTERISTICS

Almost any carbonaceous or biomass fuel can be gasified under experimental or laboratory conditions.[19] However, the real test for a good gasifier is not whether a combustible gas can be generated by burning a biomass fuel with 20 to 40% stoichiometric air but that a reliable gas producer can be made which can also be economically attractive to the customer. Towards this goal the fuel characteristics have to be evaluated and fuel processing done.

Many a gasifier manufacturer claims that a gasifier is available which can gasify any fuel. There is no such thing as a universal gasifier.[19] A gasifier is very fuel specific and it is tailored around a fuel rather than the other way round.

Thus, a gasifier fuel can be classified as good or bad according to the following parameters:

1.   Energy content of the fuel
2.   Bulk density
3.   Moisture content
4.   Dust content

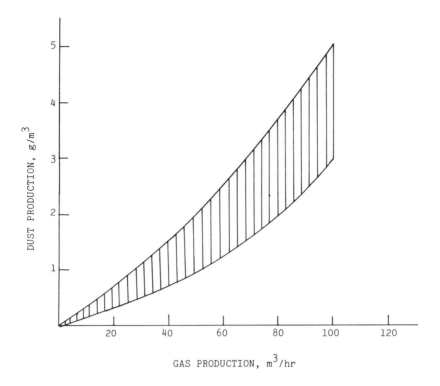

FIGURE 7.   Dust content as a function of gas production.

5.   Tar content
6.   Ash and slagging characteristic

A. Energy Content and Bulk Density of Fuel
The higher the energy content and bulk density of the fuel, the smaller is the gasifier volume since for one charge one can get power for longer time.

B. Moisture Content
In most fuels there is very little choice in moisture content since it is determined by the type of fuel, its origin, and treatment. It is desirable to use fuel with a low moisture content because heat loss due to its evaporation before gasification is considerable and the heat budget of the gasification reaction is impaired. For example, for fuel at 25°C and raw gas exit temperature from gasifier at 300°C, 2875 kJ/kg moisture must be supplied by fuel to heat and evaporate moisture.[20]

Besides impairing the gasifier heat budget, high moisture content also puts a load on cooling and filtering equipment by increasing the pressure drop across these units because of condensing liquid.

Thus, in order to reduce the moisture content of fuel some pretreatment is required. Generally, a desirable moisture content for fuel should be less than 20%.

C. Dust Content
All gasifier fuels produce dust. This dust is a nuisance since it can clog the internal combustion engine and hence has to be removed. The gasifier design should be such that it should not produce more than 2 to 6 g/m³ of dust.[19] Figure 7 shows dust produced as a function of gas production for wood generators used during World War II.[21]

The higher the amount of dust produced, the more load is put on the filters, necessitating their frequent flushing and increased maintenance.

D. Tar Content

Tar is one of the most unpleasant constituents of the gas as it tends to deposit in the carburetor and intake valves, causing sticking and troublesome operations.[22] It is a product of a highly irreversible process taking place in the pyrolysis zone. The physical property of tar depends upon temperature and heat rate and the appearance ranges from brown and watery (60% water) to black and highly viscous (7% water).[19] There are approximately 200 chemical constituents that have been identified in tar so far.

Very little research has been done in the area of removing or burning tar in the gasifier so that relatively tar-free gas comes out. Thus, the major effort has been devoted to cleaning this tar by filters and coolers.

A well-designed gasifier should put out less than 1 g/m^3 of tar.[21] Usually it is assumed that a downdraft gasifier produces less tar than other gasifiers.[25] However, because of localized inefficient processes taking place in the throat of the downdraft gasifier it does not allow the complete dissociation of tar.[19] More research effort is therefore needed in exploring the mechanism of tar breakdown in downdraft gasifiers.

E. Ash and Slagging Characteristics

The mineral content in the fuel that remains in oxidized form after complete combustion is usually called *ash*. The ash content of a fuel and the ash composition have a major impact on trouble-free operation of the gasifier.

Ash basically interferes with the gasification process in two ways:

1. It fuses together to form slag and this clinker stops or inhibits the downward flow of biomass feed.
2. Even if it does not fuse together it shelters the points in fuel where ignition is initiated and thus lowers the reaction response of the fuel.

Ash and tar removal are the two most important processes in the gasification system for its smooth operation. Various systems have been devised for ash removal.[23] In fact, some fuels with high ash content can be easily gasified if an elaborate ash removal system is installed in the gasifier.[19]

Slagging, however, can be overcome by two types of operation of the gasifier:[20]

1. Low-temperature operation that keeps the temperature well below the flow temperature of the ash.
2. High-temperature operation that keeps the temperature above the melting point of ash.

The first method is usually accomplished by steam or water injection while the latter method requires provisions for tapping the molten slag out of the oxidation zone. Each method has its advantages and disadvantages and depends on the specific fuel and the gasifier design.

Keeping in mind the above characteristics of fuel, only two fuels have been thoroughly tested and proven to be reliable. They are charcoal and wood. They were the principal fuels during World War II and the European countries had developed elaborate mechanisms of ensuring strict quality control on them.[24]

Charcoal, specifically because it is tar-free and has a relatively low ash content, was the preferred fuel during World War II and still remains so.[25] However there is a major disadvantage of charcoal in terms of energy. Charcoal is mostly produced from wood and in the conversion of wood to charcoal about 50% of the original energy is lost.[12] When made by the pit method (as is normally made in most developing countries) the losses can be as high as 80%.[7] Besides, with the present energy crisis, where most countries do not have enough supply of wood, it is advantageous and attractive to use agricultural residues. For the agricultural sector this is an extremely attractive alternative.

Many agricultural residues and fuels have therefore been gasified. However, the operating experience is very limited and most of the work has been on a laboratory scale.[15,17] Table 3 lists the characteristics of these fuels. More research needs to be done in order to make large-scale gasification systems running on these fuels.

V. GASIFICATION SYSTEMS

The combustible gases from the gasifier can be used in internal combustion engines, for direct heat applications, and as feedstock for production of chemicals like methanol.

However, in order for the gas to be used for any of the above applications it should be cleaned of tar and dust and be cooled. As previously mentioned, cooling and cleaning of the gas is one of the most important processes in the whole gasification system. The failure or the success of producer gas units depends completely on their ability to provide a clean and cool gas to the engines or for burners. Thus, the importance of cleaning and cooling sytems cannot be overemphasized.

A. Cooling and Cleaning of Gas

The temperature of gas coming out of the generator is normally between 300 and 500°C. This gas has to be cooled in order to raise its energy density. Various types of cooling equipment have been used to achieve this end.[21] Most coolers are gas-to-air heat exchangers where the cooling is done by free convection of air on the outside surface of the heat exchanger. Since the gas also contains moisture and tar, some heat exchangers provide partial scrubbing of gas.[22] Thus, ideally the gas going to an internal combustion engine should be cooled to nearly ambient temperature.

Cleaning of the gas is more tricky and is very critical. Normally, three types of filters are used in this process. They are classified as dry, moist, and wet.[22]

In the dry category are cyclone filters. They are designed according to the rate of gas production and its dust content.[26] The cyclone filters are useful for particle size of 5 μm and greater.[26] Since 60 to 65% of the producer gas contains particles above 60 μm in size, the cyclone filter is an excellent cleaning device.[21]

After passing through the cyclone filter the gas still contains fine dust, particles, and tar. It is further cleaned by passing through either a wet scrubber or dry cloth filter. In the wet scrubber the gas is washed by water in the countercurrent mode. The scrubber also acts like a cooler,[27] from where the gas goes to a cloth or cork filter for final cleaning.

Since a cloth filter is a fine filter, any condensation of water on it stops the gas flow because of an increase in the pressure drop across it. Thus in quite a number of gasification systems the hot gases are passed through the cloth filter and only then do they go to the cooler.[28] Since the gases are still above the dew point, no condensation takes place in the filter. Figure 8 shows schematically a downdraft gasification system with a cleaning and cooling train.

There is quite a substantial pressure drop across the whole gasification system and the design is usually done such that the pressure drop should not exceed 100 cm of water.[7]

B. Shaft Power Systems

The biggest application of producer gas has been in driving internal combustion engines. Both spark ignition and compression ignition engines have been driven by it. In principle any internal combustion engine can be converted to run completely or partly on the gas. However, in actual practice, running the engines uninterrupted and for long periods of time without any problem is difficult to achieve.[19]

Table 3
GASIFICATION CHARACTERISTICS OF VARIOUS FUELS

Fuel	Specifications (Treatment, bulk density, moisture content)	Tar produced (g/m³)	Ash content (%)	Gasifier	Experience	Ref.
Alfalfa straw	Cubed, 298 kg/m³, mc. = 7.9%	2.33	6	Downdraft	No slagging, some bridging	17
Bean straw	Cubed, 440 kg/m³, m.c. = 13%	1.97	10.2	Downdraft	Severe slag formation	17
Barley straw (75% straw; 25% corn fodder and 6% orza binder)	Cubed, 299 kg/m³, m.c. = 4%	0	10.3	Downdraft	Slag formation	17
Coconut shell	Crushed (1—4 cm), 435 kg/m³, m.c. = 11.8%	3	0.8	Downdraft	Excellent fuel. No slag formation	15
Coconut husks	Pieces, 2—5 cm, 65 kg/m³	Insignificant tar content	3.4	Downdraft	Slag on grate but no operational problem	15
Corn cobs	—, 304 kg/m³, m.c. = 11%	7.24	1.5	Downdraft	Excellent fuel. No slagging	17
Corn fodder	Cubed, 390 kg/m³, m.c. = 11.9%	1.43	6.1	Downdraft	Severe slagging and bridging	17
Cotton stalks	Cubed, 259 kg/m³, m.c. = 20.6%	5	17.2	Downdraft	Severe slag formation	17
Peach pits	Sundried, 474 kg/m³, m.c. = 10.9%	1.1	0.9	Downdraft	Excellent fuel. No slagging	17
Peat	Briquettes, 555 kg/m³, m. c. = 13%	—	—	Downdraft	Severe slagging	15
Prune pits	Air-dried, 514 kg/m³, m.c. = 8.2%	0	0.5	Downdraft	Excellent fuel	17
Rice hulls	Pelleted, 679 kg/m³, m.c. = 8.6%	4.32	14.9	Downdraft	Severe slagging	17
Safflower	Cubed, 203 kg/m³, m.c. = 8.9%	0.88	6.0	Downdraft	Minor slag formation	17
Sugar-cane	Cut, 2—5 cm, 52 kg/m³	Insignificant	1.6	Downdraft	Slag on hearth ring. Bridging	15
Walnut shell	Cracked, 337 kg/m³, m.c. = 8%	6.24	1.1	Downdraft	Excellent fuel. No slagging	17
Walnut shell	Pelleted	14.5	1.0	Downdraft	Good fuel	17

Table 3 (continued)
GASIFICATION CHARACTERISTICS OF VARIOUS FUELS

Fuel	Specifications (Treatment, bulk density, moisture content)	Tar produced (g/m³)	Ash content (%)	Gasifier	Experience	Ref.
Wheat straw	Cubed, 3 cm, 395 kg/m³, m.c. = 9.6%	—	9.3	Downdraft	Severe slagging, bridging. Irregular gas production	15
Wheat straw and corn stalks	Cubed (50% mix), 199 kg/m³, m.c. = 15%	0	7.4	Downdraft	Slagging	17
Wood blocks	5 cm cube, 256 kg/m³, m.c. = 5.4%	3.24	0.2	Downdraft	Excellent fuel	17
Wood chips	166 kg/m³, m.c. = 10.8%	6.24	6.26	Downdraft	Severe bridging and slagging	17

The trend at present is therefore, to use available internal combustion engines and run them on producer gas. However, since the producer gas plant is tailor made for a specific engine it is worthwhile to look at the engine itself. Producer gas being a relatively low-energy gas has certain combustion characteristics that differ markedly from gasoline or diesel oil. Thus, in future research and development in gasification it is worthwhile to do considerable work to make an engine specific for the gas. At present no such engine exists.

1. Spark Ignition Engines

When a spark ignition engine is converted to operation on producer gas it is derated to about 40 to 50%.[29] The deration is primarily because of the low energy density of producer gas. This accounts for about a 30% loss of power. The rest is accounted for by the pressure drop in the intake valves and piping.[30]

A spark ignition engine on the whole requires very little modification to run on producer gas. Generally, depending upon the make of engine (compression ratio and rpm), the ignition timing has to be advanced by about 30 to 40 degrees. This is done because of the low flame speed of producer gas as compared to gasoline.[30] The low flame speed of producer gas is more efficiently used in a low-speed engine. Thus, an engine with 1500 to 2500 rpm is ideal for producer gas applications[30] (Figure 9).

It should be noted that, in general, the overall efficiency of the internal combustion engine itself does not change, though the power derating takes place. However, detailed comparison of the engine efficiencies with and without producer gas has not been done till now because of insufficient data and large variations in producer gas composition.[30] Thus, a conservative figure of 15 to 20% can be used as the efficiency of spark ignition engines.

With the above efficiencies it is easy to calculate the mechanical energy available per kilogram of biomass gasified. With a gasifier efficiency of 68% (Section III) the total system efficiency (gasification and engine) is 10 to 13%. Thus, on average one can get 0.55 to 0.75 kWh of mechanical energy per kilogram of biomass gasified (calorific value of biomass is taken to be 19.6 MJ/kg) (Section III). This value, however, changes with the load and can go as low as 0.22 kWh/kg for 10% load to 0.83 kWh/kg for 87% load rated capacity.[31] Nevertheless, for the purpose of sizing a system a good number is 0.7 kWh/kg.

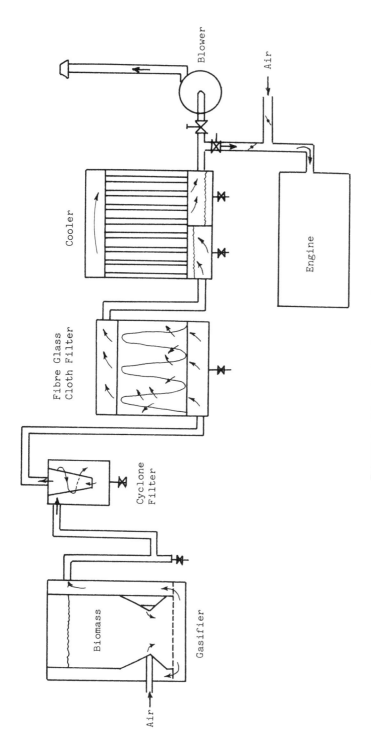

FIGURE 8.   Schematic of producer gas plant.

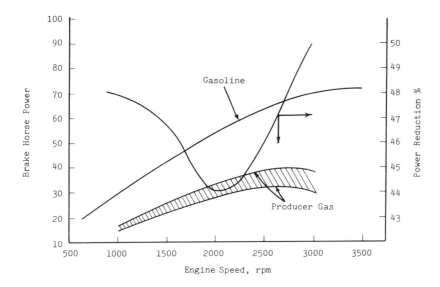

FIGURE 9. Power reduction as function of engine speed. (From Kaupp, A. and Goss, J. R., *Small Scale Gas Producer Engine Systems,* Deutsche Gesellschaft für Technische Zusammenarbeit, 1984. With permission.)

2. Compression Ignition Engine

A compression ignition or diesel engine cannot be operated on producer gas completely without injection of a small amount of diesel. This is because the producer gas cannot ignite by itself under prevailing pressure. Thus for compression ignition engines to run on producer gas they have to be either a duel fuel engine or converted into spark ignition engines.

Since diesel engines have compression ratios between 16 and 20 and are run at lower rpm than gasoline engines, they are ideally suited to run on producer gases with spark ignition. However, conversion of the engine to spark ignition is a costly and elaborate affair and the advantages are nullified by the cost.

Thus, most of the diesel engines running on producer gas have been the dual fuel type. Especially in developing countries, where the proliferation of diesel engines has been because of dual pricing structure (diesel is subsidized).

Because of the high compression ratio and low speeds, the derating of diesel engines running on producer gas is only between 15 and 30%. This is far superior to the derating of the gasoline engine. Even if gasoline engines are used in dual fuel mode, their derating is still between 40 and 50%.[30]

On the average, the diesel engine can run on 15 to 20% (of the original consumption) diesel and the rest on producer gas. Generally, the engine is started in diesel and as the gas generation builds up the diesel consumption is then kept at the idling level. The engine efficiency in this case is about 25%. Thus, as a rule of thumb, the dual fuel engine producing 1 kWh requires 1 kg of biomass and consumes 0.07 ℓ of diesel.[10]

In both the diesel and gasoline engines the introduction of producer gas to the engine is by a T valve where, from one section of the T air is sucked in. Thus the complicated carburetor is greatly simplified by the above arrangements. Many arrangements have been developed for introduction of air/gas mixture in the engine.[30]

C. Direct Heat Systems

Direct heat systems are those in which the producer gas is burned directly in the furnace or boiler. The advantage of this as compared to direct combustion of biomass is in obtaining controlled heating and higher flame temperatures than those obtainable otherwise. Because of direct burning, the gas quality is less critical than in shaft power systems and consequently they are less demanding on cooling and cleaning equipment and have more versatility as far as fuels are concerned.[7]

The direct heat systems have great attraction for agricultural applications like drying of farm produce; consequently, many such applications are underway in the U.S.[11] Since direct heat systems were rarely used during World War II, their experience is recent.

Most of the direct heat gasifiers currently available are of the updraft or fluidized bed types. Their output range varies from 0.25 to 25 GJ/hr.[7] Most of these are for large kiln and furnace applications. For agricultural drying purposes their range has to be brought down.

Since the production of gas is uneven and fluctuates with time it is sometimes necessary to have a storage of gas. This can provide very uniform quality gas. However, no such systems with storage exist at present.

Because of the low energy content of producer gas ($\sim$ 5 MJ/m^3), which is about 10 to 15% that of natural gas, special burners are needed. Since the adiabatic flame temperature of producer gas is about 1400°C,[32] the highest temperature applications can be around 1000 to 1200°C. Some manufacturers, however, have erroneously reported using the gas for 1600°C applications.[7] Most of the U.S. manufacturers producing direct heat systems have been reviewed by Goss.[11]

D. Fluidized Bed Systems

For fuels which have high ash content and the ash has low melting point, fluidized bed combustion seems to gasify them.[33]

In fluidized bed gasifiers the air is blown upwards through the biomass bed. The bed under such conditions behaves like boiling fluid and has excellent temperature uniformity and provides efficient contact between gaseous and solid phase. Generally, the heat is transferred initially by a hot bed of sand.[23] The major advantage of a fluidized bed gasifier over, say, downdraft is its flexibility with regard to feed rate and its composition. Fluidized bed systems can also have high volumetric capacity and the temperature can be easily controlled.

However, these advantages are offset by the complexity of the system with large blowers for blowing air and augers for feeding biomass. Besides, fluidized bed systems produce more dust and tar as compared to downdraft gasifiers.[33] This puts a heavy load on the cooling and cleaning train.

Nevertheless, quite a number of research projects are under way to study and optimize biomass residue-based fluidized bed gasification systems.[8,23,34] However, no large-scale manufacturing facility for such systems exists today.

VI. APPLICATIONS

As was mentioned earlier, the main applications of biomass gasifiers are

1. Shaft power systems
2. Direct heat applications
3. Chemical production

In the shaft power systems the main agricultural applications are driving farm machinery like tractors, harvesters, etc. There are quite a number of manufacturers catering to the on-farm machinery gasification systems.[36] Small-scale electricity generation systems also provide an attractive alternative to utilities.[31]

Another useful application of producer gas units is in irrigation systems. This seems to be the most important application in developing countries.[7] There is no reason why such systems cannot become popular in developed countries, especially when there have been quite a number of solar-powered irrigation systems installed.[37]

Direct heat systems, because of their simplicity, may prove to have the biggest applications in agriculture. Among them are grain drying, green house heating, and running of absorption refrigeration and cooling systems. Again these systems can be coupled to other renewable energy systems like solar for thermal applications. Another interesting application for direct heat (external combustion) application is running of stirling engines.[38] These engines have very high efficiencies and may prove to be a better alternative than internal combustion engines running on producer gas.

Production of chemicals like methanol and formic acid from producer gas is a recent phenomenon.[5] However, with fossil fuels getting scarcer, production of these chemicals by producer gas may prove to be an economically feasible proposition. Another interesting application may be the use of producer gas to run a fuel cell plant. The energy density of such a plant would be highly favourable as compared to internal combustion engine systems.

However, for all these applications the most important ingredient is the availability of biomass fuel. For on-farm applications, biomass residues are an attractive proposition. However, before any large-scale application of gasification is undertaken, the fuel availability is to be critically ascertained.

As an example, it is instructive to look at the land area required for a gasifier to run on cotton stalks (biomass residue) as fuel. On the average, the quantity of stalks harvested is 1.5 tons/acre/year.[35] Thus, a 100-kW gasifier running at 8 hr/day for 300 days/year will require about 213 acres of cotton plantation to produce the required cotton stalks. Against such background all the future applications of gasifiers should be evaluated.

If the biomass residue availability is not adequate then the decision has to be made about running it on wood. However, such decisions can only be made at specific sites and for specific applications.

Just like in any other alternative energy source, it is advisable to use hybrid systems; similarly, in biomass gasification systems it will be worthwhile to use them in conjunction with other energy systems. For example, grain drying can have biomass gasifier/solar coupling. Only in specific cases of methanol or chemical production should the gasification system be used as a separate one.

VII. CURRENT STATUS OF GASIFICATION TECHNOLOGY

An excellent survey of the current status of gasification technology has been carried out by Foley and Barnard.[41] They have reviewed the status in both developed and developing countries.

However, there is confusion regarding the number of manufacturers of gasification equipment. Quite a number of these manufacturers have just produced a few units which are still in experimental stages. There are, therefore, close to 64 manufacturers all over the world.[39,11] In the U.S. alone there are 27 manufacturers and about 13 universities and USDA research stations working on various aspects of biomass gasification.[11]

The largest gasification manufacturing facility in the world is the Gasifier and Equipment Manufacturing Corporation (GEMCOR) in the Philippines. They produce about 3000 u/year ranging in size from 10 to 250 kW. Besides, they have recently started producing gasifiers for direct heat application.[43] Their primary applications have been for irrigation pumps and power generating sets. To date about 1000 units have been installed within the Philippines running on charcoal, wood chips, and briquettes.[43] Brazil is another country where a large-scale gasification manufacturing program has been undertaken.[42] About 650 units of various sizes and applications have been installed.

In both the Brazilian and the Philippines program the gasifiers are mostly charcoal powered.

In this a strict quality control of the fuel has to be maintained. Thus, the companies involved in gasifier manufacturing also supply the quality fuel. Inadequate fuel quality is the biggest problem in running these gasifiers.[43]

In Europe there are many manufacturers, especially in Sweden, France, West Germany, and the Netherlands, who are engaged in manufacturing gasification systems for stationary applications. Most of the market for these European manufacturers has been in developing countries.[7]

The U.S. and North American manufacturing activities have been summarized by Goss.[11] In the research area the most active program in gasification is at University of California, Davis and University of Florida, Gainesville.[11] Many systems in the range of 10 to 100 kW have been developed at Davis. The U.S. also is ahead of the rest of the world in direct heat application gasifiers. Both fluidized and fixed-bed gasifiers have been developed for this purpose.

In other countries of Asia and Africa the work is being carried out in research institutions and few prototypes have been made and tested.[7] Interestingly enough there is no mention of Japan in any world-wide gasification literature. However, if the gasification technology does pick up it will be only a matter of time before Japan flexes its economic muscle and mass produces the gasifiers at cheaper rates.

Most of the gasifiers (up to the 100-kW range) being sold by different manufacturers show a leveling off price of \$380 per kW_e for plant prices and about \$150 per kW_e for basic gasifier price.[39] This leveling off comes at about the 100-kW system. However, for small systems the prices are extremely high. Thus, a 10-kW_e gasifier plant costs about \$840 per kW_e while the basic gasifier is \$350 per kW_e. To this must be added the transportation costs (especially for shipment to developing countries). These prices, therefore, can make the gasifiers uneconomic. This explains the big gasifier manufacturing push being given in countries like the Philippines, Brazil, etc.

Unfortunately, with all the activities going on around the world, the impact of gasification technology to date on the economy has been negligible and far smaller than that of other renewable energy, namely solar. However, gasification is a recently rediscovered technology and most of the development is still on the learning curve.

VIII. CONCLUSIONS

1. Biomass gasification offers a very attractive alternative energy system for agricultural purposes.
2. Most preferred fuels for gasification have been charcoal and wood. However, biomass residues are the most appropriate fuels for on-farm systems and offer the greatest challenge to researchers and gasification system manufacturers.
3. Very limited experience has been gained in gasification of biomass residues.
4. Most extensively used and researched systems have been based on downdraft gasification. However, it appears that for fuels with a high ash content, fluidized bed combustion may offer a solution. At present no reliable and economically feasible systems exist.
5. The biggest challenge in gasification systems lies in developing reliable and economically cheap cooling and cleaning trains.
6. Maximum usage of producer gas has been in driving the internal combustion engine both for agricultural as well as for automotive uses. However, direct heat applications like grain drying, etc. are very attractive for agricultural systems.
7. A spark ignition engine running on producer gas on an average produces 0.55 to 0.75 kWh of energy from 1 kg of biomass.

8. Compression ignition (diesel) engines cannot run completely on producer gas. Thus to produce 1 kWh of energy they consume 1 kg of biomass and 0.07 ℓ of diesel. Consequently, they effect 80 to 85% diesel saving.
9. Future applications like methanol production, using producer gas in fuel cell, and small-scale irrigation systems for developing countries offer the greatest potentialities.

REFERENCES

1. **Leach, G.,** *Energy and Food Production,* International Institute of Environment and Development London, 1976.
2. **Rajvanshi, A. K.,** Decentralised Technologies for Power, Indian Express, January 20, 1978.
3. **Dutta, R. and Dutt, G. S.,** Producer Gas Engines in villages of less-developed countries, *Science,* 213, 731, 1981.
4. Solar Energy Research Institute (SERI), *Generator Gas — The Swedish Experience from 1939—1945,* SERI, Golden, Colo., 1979, chap. 1.
5. **Reed, T. B., Graboski, M., and Markson, M.,** The SERI High Pressure Oxygen Gasifier, Report SERI/TP-234-1455R, Solar Energy Research Institute, Golden, Colo., Feb. 1982.
6. **Eggen, A. C. W. and Kraatz, R.,** Gasification of Solid Waste in Fixed Beds, *Mechan. Eng.,* July 1976, 24.
7. **Foley, G. and Barnard, G.,** Biomass Gasification in Developing Countries, Technical Report No. 1, Earthscan, London, 1983.
8. **Kaupp, A. and Goss, J. R.,** Technical and economical problems in the gasification of rice hulls. Physical and chemical properties, *Energy Agric.,* 1, 201, 1981—1983.
9. **Breag, G. R. and Chittenden, A. E.,** Producer Gas; Its Potential and Applications in Developing Countries, Report No. G130, Tropical Products Institute, London, October 1979.
10. **Johansson, E.,** Swedish Tests of Otto and Diesel Engines Operated on Producer Gas, Report of National Machinery Testing Institute, Sweden, 1980.
11. **Goss, J. R.,** State of Art of Agriculture Residue Gasifiers in the U.S., Proceedings First USAID/GOI Workshop on Alternative Energy Resources and Development, New Delhi, India, Nov. 7—11, 1983.
12. Solar Energy Research Institute (SERI), *Generator Gas — The Swedish Experience from 1939—1945,* SERI, Golden, Colo., 1979, Chap. 2.
13. Solar Energy Research Institute (SERI), *Generator Gas — The Swedish Experience from 1939—1945,* SERI, Golden, Colo., 1979, chap. 4.
14. **Schlapfer, P. and Tobler, J.,** *Theoretical and Practical Investigations Upon the Driving of Motor Vehicles with Wood Gas,* Bern, 1937.
15. **Hoglund, C.,** Agricultural Residues as Fuel for Producer Gas Generation, Master Thesis, Royal Institute of Technology, Sweden, 1981.
16. **Skov, N. A. and Paperworth, M. L.,** *The Pegasus Unit,* Pegasus Publishers, Olympia, Wash., 1974, chap. IX.
17. California Energy Commission, An Investigation of the Downdraft Gasification Characteristics of Agricultural and Forestry Residues; Interim Report, 1979.
18. **Ince, P. J.,** How to Estimate Recoverable Heat Energy in Wood or Bark Fuels, General Tech. Rep. FPL 29, USDA, 1979.
19. **Kaupp, A.,** Myths and Facts About Gas Producer Engine Systems, Paper presented at First International Producer Gas Conference, Colombo, Sri Lanka, 8—12 November 1982.
20. **Kaupp, A. and Goss, J. R.,** State of the Art for Small Scale (to 50 KW) Gas Producer-Engine Systems, Final Report, U.S.D.A., Forest Service, March 1981, chap. 5.
21. Solar Energy Research Institute (SERI), *Generator Gas — The Swedish Experience from 1939—1945,* SERI, Golden, Colo., 1979, chap. 5.
22. **Skov, N. A. and Paperworth, M. L.,** *The Pegasus Unit,* Pegasus Publishers, Olympia, Wash., 1974, chap. VII.
23. **O'Neill, W. and Flanigan, V. J.,** Small Fluidized Gasifier Using Charred Biomass, Presented at First International Producer Gas Conference, Colombo, Sri Lanka, 8—12 November 1982.
24. Solar Energy Research Institute (SERI), *Generator Gas — The Swedish Experience from 1939—1945,* SERI, Golden, Colo., 1979, chap. 3.
25. **Remulla, J. A.,** Gasifier Manufacture in the Philippines: Status and Prospects, Presented at Technical Consultation meeting between People's Republic of China and Philippines, Manila, June 23—30, 1982.

26. **Perry, R. H. and Chilton, C. H., Eds.,** *Chemical Engineer's Handbook,* 5th ed., McGraw-Hill, New York, 1973, 20-75.
27. GEMCOR, unpublished data, 1983.
28. Biomass Energy Consultants and Engineers (BECE), unpublished data, 1982.
29. Solar Energy Research Institute (SERI), *Generator Gas — The Swedish Experience from 1939—1945,* SERI, Golden, Colo., 1979, chap. 7.
30. **Kaupp, A. and Goss, J. R.,** State of the Art for Small Scale (to 50 KW) Gas Producer-Engine Systems, Final Report, U.S.D.A., Forest Service, March 1981, chap. 7.
31. **Van Der Heijden, S., Szladow, A. J., Barabas, M., and Sirianni, G.,** Wood Gasification system for Electricity Production, Proceedings 16th IECEC, 1981, 459.
32. Standard Handbook for Mechanical Engineers, Seventh Edition, Baumeister, T., Ed., McGraw Hill Book Co., 1967, 4-69.
33. **Maniatis, K., and Buekens, A.,** Practical Experience in Fluidized Bed Gasification of Biomass, Presented at First International Producer Gas Conference, Colombo, Sri Lanka, November 8-12, 1982.
34. **Van den Aarssen, F. G.,** Performance of Rice Husk Fuelled Fluidized Bed Pilot Plant Gasifier, Presented at First International Producer Gas Conference, Colombo, Sri Lanka, November 8-12, 1983.
35. **Rajvanshi, A. K.,** Potential of Briquettes from Farm Residues as Rural Energy Source, Proc. Workshop on Biomass Energy Management, Hyderabad, December 27-29, 1983.
36. **Hollingdale, A. C.,** Survey of Manufacturers of Gasifier Power Plant Systems, Report No. G180, Tropical Development and Research Institute, London, 1983.
37. **Barber, R. and Prigmore, D.,** Solar-Powered Heat Engines, in *Solar Energy Handbook,* Kreider, J. and Kreith, F., Eds., McGraw Hill Book Company, 1981, 22-1.
38. **Beagle, E. C.,** Gasifier — Stirling: An Innovative Concept, presented at First International Producer Gas Conference, Colombo, Sri Lanka, November 8-12, 1982.
39. **Foley, G. and Barnard, G.,** Biomass Gasification in Developing Countries, Technical Report No. 1, Earthscan, London, 1983, chap. 2.4.
40. **Baja, L.,** Personal communication, 1983.

Chapter 5

ETHANOL PRODUCTION

K. Robinson and J. Messick

TABLE OF CONTENTS

I. INTRODUCTION

Petroleum-derived fuels have been the preferred source in the U.S. market place. Their abundance and low cost were the competitive edge. With the 1970s came drastic change in the world petroleum market. In 1973 came the oil embargo and the first dramatic rise in the price of oil imposed by the Organization of Petroleum Exporting Countries (OPEC). Continued price increases and apparent shortages in the latter part of the decade started the interest in alcohol fuel as an alternative source for liquid fuel.

Alcohol fuel is emerging as a viable alternative. Because alcohol is a liquid it can be used in automobiles. Over half the liquid fuels consumed in the U.S. are used in transportation and about three fourths of that amount is used in automobiles.

Equally important is the fact that alcohol can be produced domestically from raw materials, including not only agricultural products, but wood and wood waste, coal, and even municipal garbage.

Blended with gasoline, alcohol can supplement U.S. oil usage as motor fuel extenders and octane enhancers; the near term contribution will come as a blend of 10% ethanol and 90% gasoline. Originally advertised as gasohol, these blends are not labeled or named at the filling station. Although gasohol as a product name is not seen, Figure 1 shows the steady rise in alcohol/gasoline blend sales. Figure 2 shows a corresponding rise in the production of anhydrous alcohol. The future impact of alcohol is seen by the estimated production capacity in Figure 3.

Alcohol fuel also provides business alternatives to a depressed agriculture in the U.S. Alcohol plants will compete for grain with traditional users, thus giving producers the benefit of alternative markets for their crops. Alcohol plants designed around waste products such as cull potatoes and/or fruit have the potential to utilize what was once thrown away, often at an extra expense to the producer. Plants built next to livestock operations can help in making a fully integrated system and minimizing expensive inputs.

II. HISTORY

The production of alcohol began with brewing, a part of the food supply. At the turn of the century, Henry Ford's Model T was designed to run on alcohol, gasoline, or any mixture inbetween. The agricultural base in the U.S. would have made the farmer an energy producer in the early 1900s, but alcohol fuel did not advance with the onset of increased availability of oil.

Shortages of motor fuel and rubber during World War II spurred the rapid development of alcohol fuels. Grain alcohol plants were built and whiskey distilleries were ordered to revamp to make industrial alcohol for the war effort. By 1944 the U.S. was producing close to 600 million gal of alcohol, nearly four times the production of 1942. Almost half of the 1944 production was funneled into the synthetic rubber industry. During the war, gas stations in Kansas, Nebraska, and Illinois sold an alcohol/gasoline blend called Argol. Earlier, in 1934, Hiram Walker sold an alcohol/gasoline blend which they called Alcoline.

After the war, and reduced demand for fermentation alcohol, the giant distilleries were dismantled and others were converted back into beverage plants. By 1950, less than 10% of the industrial alcohol market came from grain.

Alcohol as fuel is not a new concept and, in general, is not a new technology. Spurred largely by the interest and efforts of grass roots organizations in the latter 1970s and the desire of the American farmer, alcohol fuel began a revival. Nebraska and Iowa were the first to legislate a commission to promote the fuel. Gasohol performance tests were sponsored in a number of states. Critics were quick to find fault with the findings of most tests, but proponents were equally quick to point out the positive results and popularity of the product with motorists.

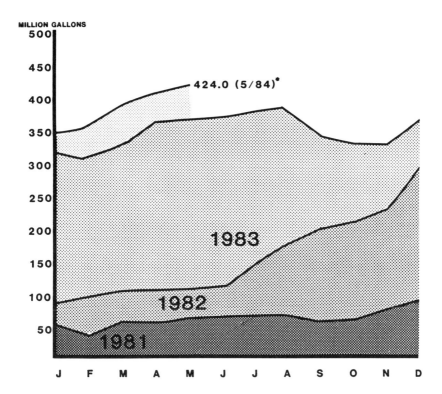

FIGURE 1. Total U.S. ethanol/gasoline blend sales: 1981, 855,264,000 gal; 1982, 2,306,100,000 gal; 1983, 4,333,904,000 gal; June 1984, 2,560,013,000 gal. (From Information Resources, Inc., *U.S. Alcohol Fuels Data Base,* Washington, D.C., 1984. With permission.)

The trade name Gasohol would eventually disappear from the market place. This can be attributed to a number of factors: during the early stages of popularity, gasoline jobbers, distributors, and handlers were not aware or were not equipped to handle alcohol. Anhydrous alcohol is necessary for blending with gasoline; a small fraction of water can cause the whole mixture to phase-separate. Isolated instances of this occurring at filling stations or in fuel tanks caused problems for the growing industry. Although "gasohol" is not seen at the pumps, the alcohol/gasoline blend market is growing.

Probably the biggest spur to the alcohol industry came when state and federal governments offered exemption for alcohol blends from gasoline excise taxes. As part of the Energy Tax Act of 1978, gasohol received a $0.04 per gallon exemption from the Federal gasoline excise tax. This translates to $0.40 per gallon of alcohol. By 1984, some 31 states exempted alcohol blends from all or part of their excise taxes, and all states had legislations pending in concern to tax exemption for alcohol blends.

Other incentives for alcohol was the Energy Security Act, which authorized $1.2 billion for loans, loan guarantees, purchase agreements, and price guarantees for the production of energy from biomass. Energy Investment Tax Credits (EITC) also provide incentives for construction of plants.

III. BASICS OF ETHANOL PRODUCTION

A. Conversion
1. Feedstocks
The extent to which alcohol production can be viable is tied directly to the raw materials available for production. Although any carbohydrate-containing material can be considered as potential feedstock, different crops or sources will be chosen as being more suitable.

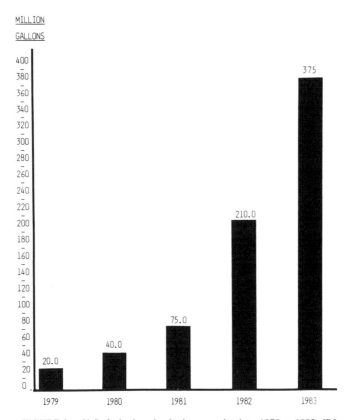

FIGURE 2.   U.S. fuel ethanol anhydrous production, 1979 to 1983. IRI second quarter 1984 estimate: 215 million gal. (From Information Resources, Inc., *U.S. Alcohol Fuels Data Base,* Washington, D.C., 1984. With permission.)

Suitability will be determined by the availability of the feedstock, yield of alcohol per ton, and in the case of agriculturally produced feedstocks, the yield of alcohol per acre.

Grain is traditionally thought of as the primary source of alcohol; in fact, a majority of the fermentation alcohol produced in the U.S. is from grain. Considering its high carbohydrate content (60 to 70%), availability in every region of the country, and with occasional surplus inventory, it is an obvious selection. Grain is not the only possibility; many other agricultural products (and their derivatives such as food wastes) and cellulosic biomass material (such as sawdust, paper, and garbage) can provide the carbohydrate needed for alcohol fermentation.

As we stated earlier, any material containing carbohydrate is a candidate for alcohol production. The carbohydrates most practical for fermentation are starch, sugar, and cellulose. All of these can be found in farm products. Table 1 shows the carbohydrate content in selected farm crops. Cellulose can be "harvested" from surplus crop residue. Crops such as corn and wheat will yield between 2.0 and 2.4 ton residue per acre (this is allowing 35% of residue to remain on soil). Table 2 shows the potential alcohol yields from various sources in the U.S. for the years 1990 and 2000.

Without a significantly high level of carbohydrate, a feedstock will probably not be considered for alcohol production. Cucumbers or melons are unlikely choices. On the other hand, consider potatoes. At first glance they show only a starch content of approximately 17%, but high tonnage yields per acre along with culls and wastes generated by commercial packaging operations make this a viable possibility.

2. Availability

In deciding the feedstock to be used in alcohol production, a detailed study should be

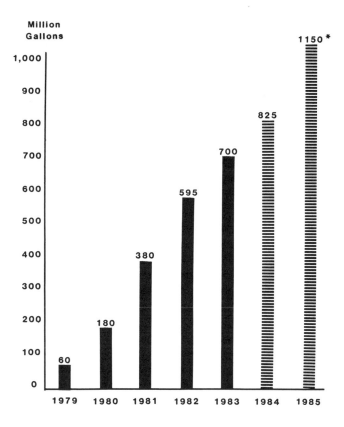

FIGURE 3. U.S. fuel ethanol production capacity 1979 to 1985. Actual operating capacity is always a varying percent of total nameplate capacity. An asterisk indicates that Government Loan Guarantees determined by facilities which finished construction by Dec. 31. (From Information Resources, Inc., *U.S. Alcohol Fuels Data Base,* Washington, D.C., 1984. With permission.)

undertaken to determine the most beneficial raw material. Factors to consider are the abundance of the raw materials as related to the expected capacity of plant production. Local crop histories should be used in this calculation. Transportation will be a key factor for larger plants. Whenever possible, reliance should not be placed on one feedstock. In any location, waste products should be considered. Most waste material will be a "product" of the food processing industry; examples include wine making, fruit canning, cheese whey, and beverage. Collection systems, seasonal or unpredictable timing, and competition from other users will be obstacles for their use.

3. Yields

The first step in calculating potential yields from any feedstock is the determination of its starch or sugar content. There are charts (Table 1) which will show the average content. For those sources where information is not available or cases where precision is required, there exist various laboratory techniques to measure fermentable carbohydrate content. Fermentable carbohydrate tests (f.c.) will "cook" the material at specific temperatures for extended periods of time in the presence of megadoses of enzyme to break all the starch into glucose. The amount of glucose is then determined by titration. Starch will break down into glucose at a ratio of just slightly over 1:1. Hydrolysis of starch will add an H_2O molecule to the resulting smaller units (glucose).

Table 1
CARBOHYDRATE CONTENT OF SELECTED FARM
PRODUCTS

Crop	Carbohydrate (%)	Crop	Carbohydrate (%)
Apples, raw	14.5	Mustard greens	5.6
Apricots, raw	12.8	Oranges	12.2
Artichokes	10.6	Peaches	9.7
Beans	64.0	Peanuts	18.6
Beets, red	9.9	Pears	15.3
Broccoli	5.9	Peas	12.0
Buckwheat	72.9	Plums, Damson	17.8
Carrots	9.7	Potatoes, raw	17.1
Cherries, sour	14.3	Pumpkin	6.5
Corn, field	72.2	Rhubarb	3.7
Corn, sweet	70.1	Rice, brown	77.4
Crabapples	17.8	Rutabagas	11.0
Cucumbers	3.4	Rye	73.4
Dates	72.9	Soybeans, dry	33.5
Dock, sheep sorrel	5.6	Squash, summer	4.2
Figs	20.3	Sweet potatoes	26.3
Grapefruit, pulp	10.6	Turnips	6.6
Grapes, American	15.7	Watermelon	6.4
Lentils	60.1	Wheat	69.1
Milk, cow	4.9	Wheat, durum	70.1
Millet	72.9	Whey	5.1
Muskmelons	7.5	Yams	23.2

From Solar Energy Research Institute, *Small Scale Fuel Alcohol Production*, Washington,
D. C., 1980.

Knowing the amount of sugar available for fermentation, we can next calculate the conversion to alcohol via fermentation. In the process:

$$\text{Glucose} \rightarrow \text{Ethanol} + \text{Carbon Dioxide}$$

$$C_6H_{12}O_6 \rightarrow 2C_2H_5OH + 2CO_2$$

The molecular weight of glucose is 180 ((6 × 12) + (12 × 1) + (6 × 16) = 180). The molecular weight of ethanol is 46 ((2 × 12) + (6 × 1) + (16 × 1) = 46). The molecular weight of carbon dioxide is 44 ((16 × 2) + (12 × 1) = 44). By balancing the equation and totaling the amount of ethanol and carbon dioxide generated can be found as:

$$C_6H_{12}O_6 \rightarrow 2C_2H_5OH + 2CO_2$$

$$180 \rightarrow 2 (46) + 2 (44)$$

$$180 \rightarrow 92 + 88$$

By weight, ethanol represents 51% of final products (92/180 = 0.511). This is not the end to calculating yields; the efficiency of fermentation must be considered along with the efficiency of operations. Fermentation efficiency will fluctuate or vary with operating conditions and types of yeast. Some sugar will be lost due to heat generation and yeast cell growth. About 94% conversion of yeast is close to the maximum efficiency for fermentation.

Table 2
BIOMASS RESOURCE BASE FOR
ALCOHOL PRODUCTION (1990—2000)

Raw material	Potential alcohol production (billions of gallons)	
	Ethanol	Methanol
Grain		
1990	4.0	
2000	4.0	
Cellulose[a]		
Wood[b]		
1990	3.2	6.2
2000	1.9	3.7
Municipal solid waste		
1990	3.7	9.8
2000	4.3	11.5
Crop wastes		
1990	1.5	3.4
2000	1.5	3.4
Subtotal for cellulose (wood, MSW, crop wastes)		
1990	8.4	19.4
2000	7.7	18.6
Sugar crops		
1990	3.0	
2000	5.0	
Sludge		
1990		0.1
2000		0.1
Food wastes		
1990	0.5	
2000	0.6	
Total		
1990	15.9	19.5
2000	17.3	18.7

[a] Cellulose may be converted to either ethanol or methanol. The figure given for each alcohol under any cellulose category is a maximum number, assuming that the particular resource is used only to produce that alcohol. A given amount of cellulose may be converted to a larger volume of methanol than ethanol. Therefore, the total volume of alcohol that may be produced from a given cellulose resource will range from the minimum (ethanol) figure to the maximum (methanol) figure.

[b] Excludes potential contribution from silvicultural energy farms.

From U.S. National Alcohol Fuels Commission, *Fuel Alcohol: An Energy Alternative for the 1980s,* Final Report, Washington, D.C., 1981.

A sample calculation for alcohol yield given 1 ton of potatoes at 16% starch:

2000 lb of potatoes × 16% starch = 320 lb total starch
320 lb starch → 320 lb sugar
320 lb sugar × 94% efficiency of fermentation = 300.8 lb available for alcohol
300.8 lb × 51% end product = 153.4 lb ethanol
153.4 lb/6.59 lb (1 gal 200° EtOH* = 6.59 lb) = 23.2 gal EtOH

* Ethanol is sometimes written in short as EtOH.

4. Preparation of Feedstocks

Preparation of feedstocks prior to cooking should accomplish two major goals: cleaning and material breakdown.

Any feedstock brought into the production facility should be inspected. Samples need to be taken for both current analysis and future references should problems arise during production. Disputes arising about the quality of feedstocks can often be settled with these samples. Material should be screened to remove contaminants such as earth, metal, or plant material (vines, roots, stems). Many grain handling devices will have magnets to remove metal parts. Aside from the fact that foreign particles will not ferment, their presence in the system can cause extensive damage to machinery like hammermills, pumps, heat exchangers, and valves.

The second goal during preparation is to physically break the material down. This will serve to expose the starch and make it available to enzyme action. In the case of sugar crops it may only be necessary to make the sugar available to the yeast. The procedure or equipment will vary from feedstock to feedstock. Two basic approaches which can be modified for many types of material are the roller-mill and hammermill.

In the roller-mill, the grain is nipped as it passes through the rollers experiencing compressive force. The roller surface may be studded or serrated to facilitate the shearing and disintegration action. In some machinery the rollers can operate at different speeds to aid in the shearing process. Variations of this machine can be used to process not only grain but crops such as Jerusalem artichokes and sugarcane.

In a hammermill the grain is delivered into a "grinding" chamber, in which a number of hammers rotate at high speed. The grain is broken down by collision with hammer and a particle wall. A retention screen will determine the final particle size by allowing only a specified particle to pass.

Alcohol yields can be affected by the grind. Higher yields wll be achieved with finer grind. A possible drawback of the fine grind is the difficulty in separating the solids from the spent mash. In cases where the stillage is being used to feed livestock, a less fine grind may be considered to facilitate the removal of solids. The trade-off will be lower alcohol yield for higher grain recovery.

5. Starch Conversion (Cellulose)

When considering "biomass" as a broad term, there are many different chemical structures to consider. For the production of ethanol through fermentation, only starch and cellulose should be considered.

In the case of grains and tubers, starch is being broken down. Starch is chemically composed of two types of long chains of glucose units: amylopectin and amylose. Amylopectin is multibranched with 40 to 70 chains of about 25 glucose units each. Amylopectin accounts for 75 to 85% of most starches. Amylose is a linear chain of about 200 to 1000 glucose units each (see Figures 4 and 5).

These molecules are very large and are responsible for the viscosity of starch solutions and pastes, as well as its characteristics of gelatinizing when heated in water.

Enzymes, capable of degrading starch, are widely distributed. These enzymes all act by breaking the linkage between adjacent glucose units, specifically hydrolysis of α-1,4 and/or α-1,6 glucosidic linkages. The enzymes needed to perform these tasks fall into four groups (see Table 3). The four groups are the α-amylases, which cause end-cleavage of substrate; β-amylases, which hydrolyze alternate bonds from the nonreducing end of the substrate; amyloglucosidases, which hydrolyze successive bonds from the nonreducing end of the substrate; and the debranching enzymes, which cleave the α-1,6 glucosidic linkages. Table 4 shows the characteristics of these enzymes. β-Amylase has the ability to produce maltose rapidly. The enzyme degrades amylose, amylopectin, or glycogen in an exo-or step-wise

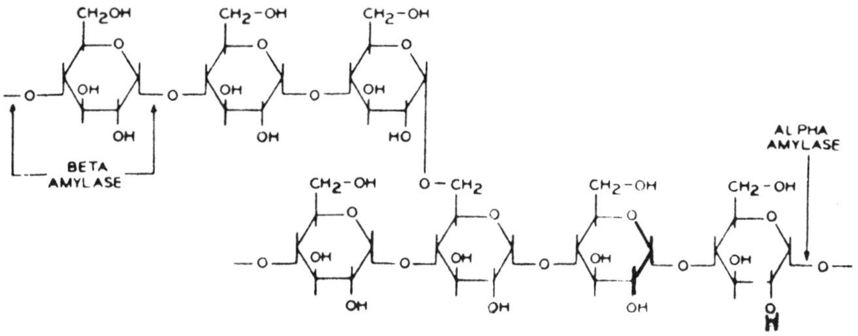

FIGURE 4.   Amylopectin molecule.

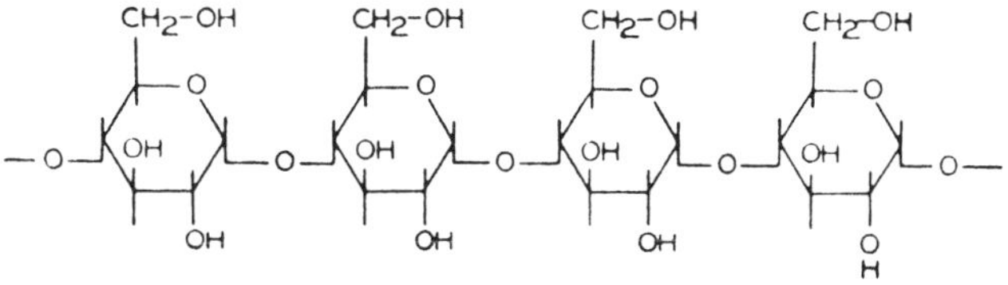

FIGURE 5.   Amylose molecule.

Table 3
COMMERCIALLY IMPORTANT STARCH-DEGRADING ENZYMES[4]

Trivial name	Systematic of scientific name
α-Amylase	α-1,4-glucan 4-glucanohydrolase
β-Amylase	α-1,4-glucan maltohydrolase
Amyloglucosidase	α-1,4-glucan glucohydrolase
Debranching enzyme	α-1,6-glucan 6-glucohydrolase

Table 4
CHARACTERISTICS OF STARCH-DEGRADING ENZYMES[4]

	α-Amylase	β-Amylase	Amyloglucosidase	Debranching enzyme
Hydrolyze α-1,4-glucosidic bonds	Yes	Yes	Yes	No
Hydrolyze α-1,6-glucosidic bonds	No	No	Yes	Yes
Ability to by-pass α-1,6 branch points	Yes	No	Bonds cleaved	Bonds cleaved
Configuration of C_1 of product	α	β	β	—
Mechanism of substrate attack	Endo	Exo	Exo	—
Viscosity reduction relative to bond hydrolysis	Fast	Slow	Slow	—
Production of reducing sugars	Slow	Fast	Fast	—
Iodine staining power	Decreased	Decreased	Decreased	Increased

Table 5
CLASSIFICATION OF DEBRANCHING ENZYMES[4]

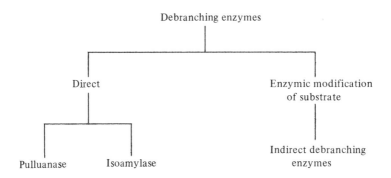

fashion by hydrolyzing alternate glucosidic bonds. Degradation takes place with invasion of configuration and therefore β-amylase is incapable of by-passing branch points, i.e., α-1,6-glucosidic linkages in amylopectin and glycogen.

This action results in the conversion of approximately 50 to 55% of amylopectin maltose; the remainder is a large limit dextrin.

The debranching enzymes fall into two categories: direct and indirect (see Table 5). Direct debranching enzymes will degrade immodified amylopectin or glycogen. The substrate polymer must first be modified by another enzyme or enzymes before an indirect debranching enzyme will act on the polymer. Amyloglucosidase, although not specifically a debranching enzyme, serves that function during the conversion process. Amyloglucosidase (glucoamylase) converts amylopectin and glycogen almost quantitatively into D-glucose. The actual specificity of amyloglucosidase appears to be in doubt, as the enzyme cleaves α-1,3 and α-1,6 as well as α-1,4 linkages.

With an understanding of the role of such enzymes, a process becomes apparent which allows the conversion of starch to fermentable sugar. After the product is milled, water and α-amylase are added to the mixture and the solution is heated to boiling.

The α-amylase will break down the starch to maltose and limit dextrins, although complete breakdown will not be accomplished. The main role of α-amylase during the heating stage is to act as a liquifier (hence the name "liquifying enzyme") and eliminate the viscosity lumps that occur during cooking (62 and 92°C for most starches, except for wheat, rice, and potato starch, which do not lump at 92°C). After boiling, the solution is allowed to cool, at which time α-amylase is added a second time to continue the breakdown of 1,4 linkages (boiling temperatures have destroyed the first addition). When the mixture has cooled to about 45 to 50°C a second enzyme is added. This enzyme is usually not one, but a mixture of enzymes including α- and β-amylases and amyloglucosidase. These enzymes will remain active through the cooking and into the fermentation stage, where amylose and amylopectin will continue to break down into fermentable sugars. The use of a debranching enzyme is not needed with the combined activity of α- and β-amylase and amyloglucosidase.

In any process designed for alcohol production using enzymes, consideration has to be given to the optimum operating conditions. Enzymes will have activity ranges, controlled by such factors as temperature, pH, calcium ion concentration, and substrate makeup. These activity ranges will vary from enzyme to enzyme and sometimes from manufacturer to manufacturer. The design and operating procedure of any plant will have to take these activity ranges into consideration to insure complete breakdown of starch (or feedstock) into fermentable sugar. An example would be commercial amyloglucosidase which has an optimum temperature for conversion of 60°C and a pH of 4.0 to 4.5. Holding the mash and enzymes under these conditions will insure the hydrolysis of the greatest number of bonds per minute. The amount of time to hold the mixture under these conditions will depend on the amount of starch and the amount of enzyme.

Table 2 should give indication to the potential of alcohol production from waste, most of which is in the form of cellulose. Many scenerios could be drawn about the utilization of municipal solid wastes, agricultural wastes, animal wastes, and others and the subsequent impact on the supply of fuel even considering the cost of collection and transportation. As promising as these figures may seem, there are problems in the breakdown of cellulose into usable components (not necessarily fermentable sugar).

There are two major obstacles when considering the hydrolysis of cellulose. First, the insolubility and crystalline structure of cellulose, which causes the reaction rate to depend on the available surface. Second, many cellulose wastes contain lignin which does not chemically inhibit cellulose, but may limit the access of the cellulose to the enzyme.

Cellulosic solids will usually contain three major components: cellulose, hemicellulose, and lignin. Cellulose is an homogeneous polymer of glucose; hemicellulose molecules are often polymers of pentoses, hexoses, and a number of sugar acids. Lignin is a polyphenolic macromolecule.

Unlike hemicellulose, cellulose is strongly resistant to hydrolysis, both enzymatic and acidic. Cellulose is a linear glucose molecule with 1,4-β-glucosidic linkages. Technically, the 1,4-β-glucosidic linkage is no harder to break than the 1,4-α-glucosidic bonds found in the starch molecule if the bond is fully exposed and free from lignin. The difficulty is not the linkage, but the structure of the cellulosic raw material.

There are two factors which make degradation of cellulose, enzymatic or acidic, difficult to achieve. Acids are nonspecific catalysts, attacking lignin as well as cellulose. An acid such as sulfuric, will delignate and also work as a catalyst for cellulose hydrolysis, causing only amorphous cellulose to hydrolyze. (The easily hydrolyzed portion of cellulose is referred to as the "amorphous" region, the resistant portion is the crystalline cellulose.) On the average, cellulose is 15% amorphous and 85% crystalline. The 15% amorphous regions could be the area of the crystallite where the molecule is "bending" or "folding" back. This exposes the β-linkages, making them accessible to cleavage.) Because of strong crystallinity, a strong acid is needed to complete a high percent of conversion.

Cellulose will provide untapped reserves for future use as an energy source. Its use as a raw material will face a number of difficulties: (1) pretreatment — wastes residues are not susceptible to either enzyme or acid hydrolysis and therefore will require some form of pretreatment; (2) pretreatment can be extremely expensive in terms of energy and cost; (3) strong acid for hydrolysis is expensive, dangerous, breaks down glucose, reduces yields, and corrodes equipment; (4) enzyme technology has not achieved low-cost production of the cellulose enzymes.

Currently, the production of alcohol from cellulose is not commercially attractive. Logistic and technical barriers must be overcome. As interest in cellulose increases and our energy supplies dwindle, advances in cellulose technology can be expected.

B. Fermentation

Chronologically, this stage of the alcohol process will take place directly after cooking or conversion. Fermentation is a major step as it is here that yeast produces alcohol from fermentable sugars. Two types of fermentation processes exist today: batch fermentation and continuous fermentation. Batch fermentation is the traditional commercial approach to alcohol production.

The typical procedure for a batch fermentation begins with the filling of the fermentation vessel with mash or wort (industry name referring to the sugar solution resulting from the

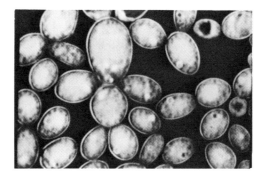

FIGURE 6.   Budding yeast cells.

cooking stage) and then cooling to the proper temperature. Optimum temperature for the production of alcohol from yeast is 25 to 30°C. Brewery fermentations may be much cooler than this. (Note that a brewery will be concerned with other factors besides maximum alcohol production such as flavor, taste, and texture.) The sugar concentration of the wort will depend greatly on the feedstock being used. This concentration can be measured with a Brix hydrometer, which will measure specific gravity and relate this to sugar concentration. Ideal concentrations for fermentation will be between 16 and 25° Brix. While checking the temperature and Brix of the wort, it is important to check the pH of the solution and adjust, if needed, to 5.5. Yeast is now added to the wort, usually in a slurry form; this process is usually referred to as "pitching" the yeast. Fermentation starts slowly and then proceeds vigorously for approximately 30 hr. Fermentation activity can be measured by the decrease in the Brix reading. If a dried or pressed yeast is used it may be 10 to 12 hr before a decrease in Brix is noticed. When active fermentation begins the Brix will drop at the rate of 0.1° per hour accompanied by evolution of heat.

$$C_6H_{12}O_6 \text{ yields } 2C_2H_5OH + 2CO_2 + 2ATP + 56 \text{ kcal}$$

The final Brix should be between 2 to 5°. With the drop in Brix will come a rise in alcohol content. Normal fermentations should be over within 48 hr with a final alcohol content determined by the beginning sugar concentration.

1. Yeast

Yeast is the microorganism that will bring about the transformation of sugar to alcohol. Yeasts are unicellular fungi. Many types of yeast can ferment sugar and they are grouped under the genus *Saccharomyces,* meaning "sugar splitting". The species which has the greatest capacity to ferment sugar to alcohol is *S. cerevisiae* (Figure 6 shows a microscopic view of budding yeast) and within this species hundreds of variants have been identified.

The normal cycle of a yeast fermentation goes through several phases: (1) lag phase; (2) exponential growth phase; (3) deceleration phase; (4) stationary or resting phase.

The first phase, the lag phase, is the period after "pitching" in which the yeast is adapting to the solution. In this period, the yeast cell will adjust to the osmotic pressure, the pH of solution, the temperature of the solution, and will then begin to feed on some of the available nutrients. The length of time taken up during this phase will depend mostly on the form the yeasts were in when pitched. For example, dry yeast or even cake yeast will need extra time to go into suspension and to begin metabolic action, while a "starter culture" of yeast may have already accomplished this. In most commercial facilities, a starter culture is used. This is usually done by taking a smaller vessel and adding the nutrients, mash, and yeast and then allowing the wort to ferment for 6 to 10 hr. This mixture is then added to the fermentation tank.

Following the lag phase the yeast moves into the exponential growth phase and multiplies. Yeast reproduces by budding daughter cells from parent cells. This reproduction will continue as long as the solution contains sufficient oxygen and nutrients. For maximum reproduction the solution requires: (1) adequate oxygen supply; (2) adequate sugar supply; (3) adequate nitrogen supply; (4) adequate mineral supply; (5) proper pH and temperature.

When the dissolved oxygen is depleted, the yeast enters a period of negative acceleration. This is the period in which alcohol fermentation occurs, provided adequate nutrients are available in the solution.

The final stage sets in toward the end of fermentation. The yeasts are still active although at a reduced rate. The availability of sugar has decreased and the concentration of alcohol has risen, both of which lead to the slowdown of fermentation.

During the first two phases of fermentation, the yeast cell population increases dramatically, from inoculation rates of 1×10^7 to almost 1.8×10^8 yeast cells per milliliter at the end of the exponential growth phase. The peak cell population will correspond to oxygen depletion in the fermenter. The metabolic reactions of the yeast change to anaerobic (non-oxygen requiring) from aerobic (oxygen requiring). The result of the anaerobic stage is the production of alcohol. The production of ethanol from yeasts is the result of its searching for an alternative hydrogen acceptor to oxygen. Any action during this phase of fermentation that might increase the oxygen content will decrease the amount of alcohol produced and lengthen the fermentation time.

We can identify two cycles occurring during the fermentation process: aerobic respiration and anaerobic respiration.

Aerobic Respiration:
Glucose $+ O_2 \rightarrow$ Yeast $+ CO_2 + H_2O$

Anaerobic Respiration:
Glucose $+$ Yeast $\rightarrow$ Ethanol (EtOH) $+$ Yeast $+ CO_2$

The anaerobic stage can be viewed as this:
Glucose $\rightarrow$ Alcohol $+$ Carbon Dioxide $+$ Heat
$C_6H_{12}O_6 \rightarrow 2(C_2H_5OH) + 2CO_2 +$ Heat

When considering an alcohol production process, other considerations to be aware of for both design and operation of the facility are temperature, inoculation, infection, and carbon dioxide.

2. Temperature

Remember that fermentation is a biological process and that temperature can affect the process. Fermentation rates will increase with rise in temperature, with the optimum temperature being between 85 and 104°F. Different strains of yeast will vary for their own optimum temperature. Most fermentations will be run at temperatures between 90 and 95°F. Lowering the temperature below this range will just lengthen the fermentation time; raising the temperature above this can run the risk of stalling the fermentation, that is operating at a temperature where the yeasts are unable to function, thus slowing or completely stopping fermentation. Exposure of yeast to 140°F for 1 min will kill the yeast cells; exposure to lower temperature for longer periods of time can also kill yeast. Allowing the fermentation temperature to raise above 104°F may not kill the yeast but may force inoculation with fresh yeast in order to complete the fermentation.

The fermentation process is heat-generating or exothermic. A 20% sugar solution can theoretically raise the temperature by 95°F. Although approximately one third of this heat will be lost to evaporation and other means, some 60°F must be removed during the process. Cooling can be accomplished in a number of ways. The best method will depend on fermenter design and the characteristics of the brew. Internal cooling can be with coils or panels using a coolant or cold water to recirculate. External cooling can be with a cooling jacket, again using coolant or water as the cooling agent, or heat exchangers. The brew is removed from the fermenter and passed through a heat exchanger and then back to the fermenter. This can be energy demanding and increases the likelihood of infection.

3. Inoculation

The rate of fermentation is directly related to the yeast inoculation rates. Higher yeast populations can decrease fermentation time. Normal batch fermentations will have yeast populations of 120 to 160 million cells per milliliter whereas continuous fermenters might have ten times the population and likewise cut fermentation times down to 4 to 6 hr. Inoculation rates of 5 to 10 million cells per milliliter in a batch fermentation will result in a final population close to 120 to 160 million yeast cells per milliliter. This rate can be accomplished by adding yeast slurry at 3 to 4% by volume of mash, pressed yeast at 12 lb/1000 gal, or dried yeast at 4 lb/1000 gal.

4. Infection

Infection is always a problem in alcohol fuel production. Fuel alcohol plans usually lack the sophistication of breweries and distilleries, which provide for adequate cleaning of all tanks and lines and efficient cooking and cooling of mash and by-products.

Infection in the mash can lead to a number of problems:

1. Spoilage organisms can compete with yeast for available sugar, thus lowering the alcohol yield.
2. Spoilage organisms may have metabolic products which inhibit the yeast.
3. Spoilage organisms may be capable of utilizing ethanol in their metabolism.

Most wort or beer spoilage organisms are killed in the cooking operation; however, there is a thermophillic spore-forming bacteria which will survive boiling. *Termobacteria*, family Achromobacteriaceae, impart a putrefactive odor and taste to beer. Their presence is very common and can be traced to the original grain or the dust in the air. *Termobacteria* can normally be killed during the growth phase of fermentation by the dominating action of yeast. This is an important reason to promote a vigorous and quick fermentation.

Acetic acid bacteria of the *Acetobactor* genus are a common beer spoiler. These bacteria ferment ethanol to acetic acid, and are aerobic, requiring oxygen to survive. They will not usually survive an anaerobic fermentation. They can infect the beer after fermentation if the beer is inoculated by an unclean surface. Closed tanks with head space filled with CO_2 will help to prevent spoilage by acetic acid bacteria. Distillation soon after the finish of fermentation is a good insurance measure.

Lactic acid bacteria of the Lactobactyeriaceae family and *Pediococcus* genus are a serious threat to the spoilage of beer and also a contaminant of the yeast. They are generally anaerobic but to a degree facultative anaerobic. These bacteria ferment sugar to lactic acid even if small traces of sugar are present. They propagate rapidly under similar conditions to yeast growth. Yeast can overpower them in fermentation, but lactic acid bacteria is a very difficult organism to control and requires regular cleaning of the fermenter tanks to prevent the bacteria from growing.

Wild yeast can be a spoiler of the fermentation where they contaminate the in-house yeast

FIGURE 7. One idea for facilitating cleaning is to make tanks that have sloped bottoms for easy draining (Colorado Agro-Energy Inc. plant under construction, Monte Vista). (From Solar Energy Research Institute, *A Guide to Commercial Scale Ethanol Production*, Washington, D.C., 1981.)

and eventually dominate the yeast culture. Wild yeast are best detected by microscopic examination. Wild yeast are usually elongate or sausage-shaped cells. Routine cleaning of fermenters and culture tanks is vital to controlling wild yeast contamination. In the event that wild yeast enter or dominate the culture yeast, sluggish fermentations and poorly alternated fermentations will result. Discarding the yeast and thorough cleaning is the only remedy.

Detection of infection can be by one of these methods:

1. Microscopic examination
2. Monitoring of acid levels (pH) during fermentation
3. Presence of unusual odors signaling spoilage products
4. Slow or sluggish fermentations

Broadly, infections can be prevented by thorough cleaning techniques, active and vigorous fermentations, avoiding holding times during any stage of the process, and quality control monitoring during the process. Figure 7 shows the sloped bottom foundation of fermentation tanks. Draining and cleaning will be aided by this design.

C. Distillation

1. Alcohol Recovery and Dehydration

The objectives in the alcohol recovery and dehydration areas of a fuel alcohol plant are twofold: recover the alcohol from beer and distill the alcohol to a refined product containing less than 0.4 wt % water (about 199° U.S. Proof). The alcohol is recovered in a stripper-rectifier (beer still) as an overhead product. Because alcohol and water form an azeotrope with a composition of about 96 wt % alcohol, the stripper-rectifier produces an overhead alcohol product of about 190° U.S. Proof. The hydrous alcohol is dehydrated in an azeotropic distillation system. Figure 8 shows the vapor/liquid equilibrium diagram that characterizes the binary azeotrope of several azeotropes.

A detailed discussion on azeotrope distillation technology is beyond the scope of this book. In general, the technology is based on the addition of a third component to the binary

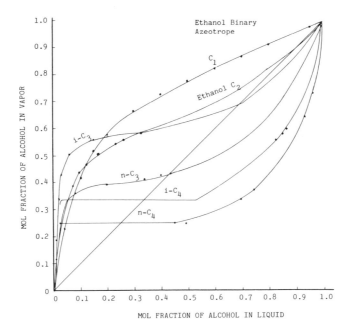

FIGURE 8. Vapor-liquid equilibrium data alcohols and water. (From Perry, R. H. and Chilton, C. H., *Chemical Engineers' Handbook,* 5th ed., McGraw-Hill, New York, 1973. With permission.)

mixture of alcohol and water. The third component is called an azeotrope agent. The agent must be a heterogenetic azeotrope, i.e., the liquid condensate from the overhead vapor of the distillation column will separate into two phases. Figure 9 is the ternary phase diagram of a typical azeotropic agent.

Because of the cost of energy, it is important for the process to be energy efficient. An energy-efficient design is more complicated than a design that uses considerable energy. Some of the standard energy-saving features of a modern design include:

- Preheat the incoming beer with overhead vapors from the dehydration tower or stripper-rectifier
- Pressurization of the dehydration tower or the stripper-rectifier such that the overhead vapors can be used to boil-up the tower that is not pressurized
- Flash the bottoms from the stripper-rectifier to preheat the beer
- Thermocompress the stripper-rectifier bottoms for steam addition to the stripper-rectifier
- Pressurize the hydrocarbon stripper for preheating the beer, or boil-up of the stripper-rectifier or dehydration tower

These features can decrease the energy usage from 40 lb of steam per gallon of product for a simple design to a system that uses only 15 lb of steam per gallon of product for a modern, energy-efficient design.

An example of a modern design is represented by Figure 10. The stripper-rectifier is pressurized to preheat beer and provide heat to the dehydration tower and the hydrocarbon (azeotrope agent) stripper. The bottoms from the stripper-rectifier are flashed for heat recovery to preheat beer. Overhead vapors from the dehydration tower and hydrocarbon stripper are used to preheat beer. Energy usage for this design is about 18 lb of process steam per gallon of product. An explanation of the process for recovery and dehydration of alcohol from a grain dry-milling plant is described below.

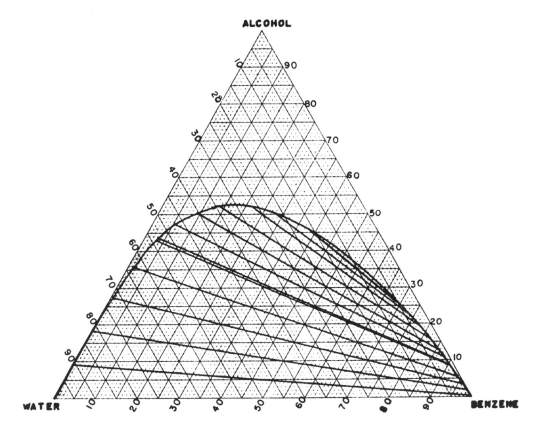

FIGURE 9.   Benzene-alcohol-water at 25°.

The beer feed contains about 7.1 wt % alcohol and 7.0% solids. Solids consist of both soluble and suspended solids. These occur in approximately equal amounts. The beer leaves the beer well at a temperature of 90°F and undergoes a series of preheating steps before it enters the first stage of distillation. The beer first passes into the tube side of a preheater. In this unit, approximately 23% of the total preheating is accomplished. The first preheating step utilizes a portion of the vapors from the dehydration tower. These vapors are condensed to supply this first stage preheating. The warmed beer feed next passes to the condenser-preheater where additional preheat, amounting to about 8.5% of the total, is added. In this condenser-preheater, a portion of the overhead vapors from the pressure stripper-rectifier is condensed to supply second stage preheat. The beer feed next passes through two stages of feed preheating, wherein a portion of the heat in the bottom stream from the pressure stripper-rectifier tower is utilized in a two-stage flash operation. These stages add about 21.5% of the total feed preheating. The warm beer next passes into a steam condenser where low-pressure steam is used to accomplish additional preheating. Approximately 23% of the total feed preheat is added in the steam condenser. The heating medium in this case consists of low-pressure steam taken from other parts of the plant. The feed is finally preheated, approximately to saturation temperature, in an additional two-stage heating by use of the flash heat taken from the bottom stream out of the pressure stripper-rectifier. Approximately 24% of the total feed preheating is accomplished in the additional two stage heating.

The hot, saturated, dilute beer feed next passes into a degassing drum where carbon dioxide dissolved in the beer feed is flashed off. The carbon dioxide dissolved in the beer feed is flashed off. The carbon dioxide represents one of the by-products of the fermentation reaction. Any alcohol or water vapor accompanying the vented carbon dioxide is condensed and drained to a flash drum.

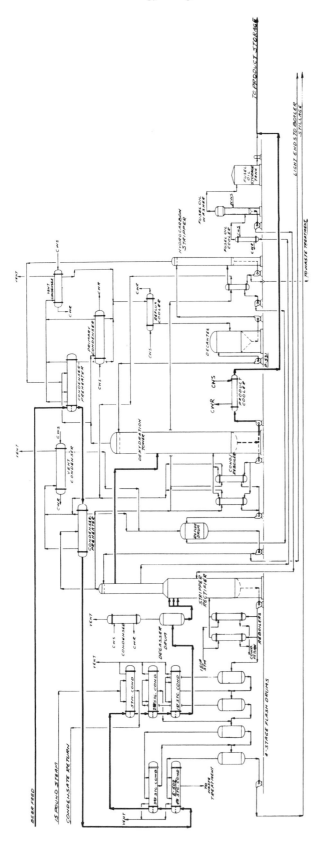

FIGURE 10.   Modern design of a stripper-rectifier.

The saturated dilute beer feed enters the midsection of the stripper-rectifier. Because of the high suspended solids content of the beer feed, the lower section of the stripper-rectifier has been designed as a disc-and-donut-type tower. This represents an effective contacting device which tends to be self-purging, and does not allow the buildup of solids which would block ordinary distillation trays. The stripper-rectifier operates with a head pressure of 50 psig. The nonvolatile soluble solids and suspended solids in the dilute beer feed wash down through the stripping section of the stripper-rectifier and a very dilute alcohol stream, containing less than 0.02 wt % alcohol, is removed from the bottom of the tower. The dilute stillage containing the dissolved and suspended solids leaves the base of the tower at about 304°F. In the bottom section of the tower, the alcohol is effectively stripped from the dilute beer. The aqueous bottom stream then passes through a series of flash stages. These stillage bottoms are subjected to progressive reductions in pressure through four flash stages. The flash vapor that develops in these stages is utilized to accomplish a portion of the feed preheating, as described previously. In these four flash stages, the temperature of the hot stillage is reduced from 304 to approximately 212°F.

Heat is supplied to the base of the stripper-rectifier by means of condensing 150 psig steam on the shell side of parallel forced-circulation reboilers. Total steam supplied to the base of the tower through the shell sides of the reboilers is 18 lb/gal of product.

Alcohol-rich vapors generated in the pressure stripper-rectifier pass overhead from the tower at a temperature of about 250°F and a pressure of 50 psig. These vapors may be utilized as a source of heat by condensing in the reboilers which are attached to the base of the dehydration tower and the hydrocarbon stripper. Sufficient vapor is generated in the pressure stripper-rectifier to allow a portion of the total overhead vapor to be utilized in a condenser-preheater to do some of the feed preheating which has been described previously. Of the total overhead vapor generated in the pressure stripper-rectifier, 10.9% is utilized for feed preheating, 81.6% is used to supply heat to the dehydration tower, and 7.5% is employed to supply heat to the hydrocarbon stripper. The overhead vapor from the pressure stripper-rectifier contains 95 vol % alcohol (190° proof).

The upper five trays of the pressure stripper-rectifier operate in a total reflux condition. The liquid product from the pressure stripper-rectifier is removed as a liquid side draw stream about five trays from the top of the tower. From here, it passes to the midsection of the dehydration tower.

The dehydration tower operates at essentially atmospheric pressure. The bottoms stream from the dehydration tower represents the anhydrous motor fuel-grade alcohol and has a concentration of 99.5 vol % ethanol (199° proof), the balance being water. The bottoms stream from the dehydration tower is pumped through a product cooler to reduce the temperature of the product alcohol to about 90°F.

Heat is supplied to the base of the dehydration tower through parallel forced-circulation reboilers. The overhead product from the dehydration tower is a ternary minimum boiling azeotrope consisting of hydrocarbon, alcohol, and water. A portion of these overhead vapors is utilized for feed preheating and the balance of these vapors is condensed. The condensed vapors pass to a reflux cooler where they are further subcooled. The subcooled liquid then passes to a decanter.

The subcooled liquid entering the decanter separates into two layers. The upper layer is the largest in volume and represents the hydrocarbon-rich layer. The lower layer, which separates, is a water layer containing some alcohol and hydrocarbon. The upper layer from the decanter is pumped back to the top tray of the dehydration tower.

The lower layer from the decanter is pumped to the top tray of the hydrocarbon stripper. The hydrocarbon stripper serves to strip the remnants of hydrocarbon and alcohol contained in the feed to the top tray. The bottom stream from the hydrocarbon stripper is an essentially aqueous stream. Thermal energy is supplied to the base of hydrocarbon stripper via alcohol-rich vapor condensing on the shell side of the reboiler. The condensed alcohol-rich vapor is collected in a reflux drum, where it joins other vapor condensate before being returned to the top tray of the pressure stripper-rectifier. Overhead vapors from the hydrocarbon stripper, containing alcohol, hydrocarbon, and water, pass to a condenser-preheater where they are condensed. The condensate is returned through a reflux cooler to the decanter. The aqueous stream passing from the bottom of the hydrocarbon stripper contains less than 0.02 wt % alcohol and hydrocarbon.

In the yeast fermentation process, certain extraneous products, in addition to ethyl alcohol, are formed. These are generally higher alcohols, i.e., higher molecular weight alcohols which are known as fusel oils, and light ends which include such things as aldehydes, etc.

The distillation system provides for the removal of these extraneous components in the following way. The higher alcohols, or fusel oils, have the property of being more volatile than alcohol in dilute aqueous solution, but they are less volatile than alcohol in concentrated alcohol solution. For this reason, they tend to concentrate on some tray in the rectifying section of the pressure stripper-rectifier. These fusel oils, thus having concentrated, can be removed as a liquid side draw stream from the pressure stripper-rectifier. They are removed and passed through a fuel oil cooler. From there, they pass to a fusel oil washer. A heavy aqueous stream containing the extracted ethyl alcohol is removed from the base of the fusel oil washer. This stream is returned to the lower section of the pressure stripper-rectifier for alcohol recovery. The light fusel oil stream is decanted from the top of the fusel oil washer and passes on to the fusel oil storage tank.

In general, the fermentation process when utilizing corn will produce about 4 to 5 gal of fusel oil for every 1000 gal of anhydrous alcohol product. These fusel oils have a heating value and can be reblended into the product. If this reblending operation is not desired, then the fusel oils may be passed to the plant boiler where they are used as fuel. Fusel oils should have no harmful effect upon fuel-grade alcohol.

Light extraneous products from fermentation, such as aldehydes, are effectively removed in this distillation system by withdrawing a very small purge from the total reflux stream passing back to the top tray of the pressure stripper-rectifier. This light component purge, in general, cannot be reblended into the alcohol which is to be used for fuel blending. The problem is that these would tend to cause vapor lock when this material is blended with gasoline. Therefore, these materials are removed and sent to the plant boiler where their fuel value is recovered.

The fusel oil and light ends must be removed because their presence would upset the equilibrium associated with the dehydration step, and could bring about problems in the decantation step occurring in the decanter.

The distillation scheme* for producing motor fuel-grade alcohol, as depicted in Figure 10, utilizes only 18 lb of process steam per gallon of anhydrous motor fuel-grade alcohol product. This great reduction in energy use is accomplished by optimizing the feed preheating scheme and by utilizing the heat content of high-pressure vapors, produced in the pressure stripper-rectifier, to supply the reboil heat for both the dehydration step and the hydrocarbon stripping step.

D. By-Products

An important part of the overall economics of an alcohol plant is the use of by-products along with the alcohol produced. The two major by-products found in the typical alcohol plant are distillers' grains and carbon dioxide.

During the anaerobic fermentation of sugar to ethanol, an almost equal amount of CO_2 to EtOH is produced. In order to collect the CO_2, closed vessels must be used and the gas

---

* Patent by Raphael Katzen Associates International, Inc.

generated must be pumped through scrubbers to remove contaminants. The gas is then generally dried and compressed.

The decision to collect CO_2 may be different for each plant. Size is the first factor to consider. A plant will probably be producing 6 to 10 million gal of ethanol per year before it begins to look feasible for collecting CO_2. Other considerations will be the capital investment for collection. In some cases, companies interested in purchasing CO_2 will also be prepared to collect and process the gas on site. Naturally their factors will be the local demand for CO_2 and local market price.

Stillage is the by-product of distillation. It is what remains of the beer after all the alcohol has been removed. Basically, the stillage will contain all the original proteins and vitamins present in the feedstock as the starch or sugar was removed in the production of alcohol. The exact makeup of the stillage will depend on the feedstock used and the process used in the plant. Most information relates to residue from feedstocks such as corn, milo, wheat, barley, and rye. Very little information is available on stillage from such feedstocks such as potatoes, sugar beets, and sugar cane.

Whole stillage will typically average 5 to 10% dry matter (DM). This concentration will depend on several factors:

1. Type of feedstock
2. Water to grain ratio used in cooking
3. Efficiency of fermentation

Feeding this material to livestock has met with varying degrees of success. The high water content presents problems in transport and storage costs. The material also offers an excellent media for bacteria growth; thus, feeding or further processing cannot be delayed long. Reduced performance may also result with feeding due to the large amounts of water which must be consumed to get the desired nutrients. The benefits of feeding whole stillage are the removal of the high energy costs associated with drying and handling of the material. A situation which provides for immediate feeding of stillage on site can offer positive economics to a facility.

When the usage of whole stillage is not practical or possible, the coarse grain is removed from the whole stillage with a screen and press or a centrifuge. These grains will range from 50 to 70% DM depending on the process used. This material is then sent on to drying equipment. The thin stillage which contains the yeast cells and other soluble nutrients can then be condensed by use of an evaporator. The evaporator will give us a material ranging in DM content from 20 to 40%. These solubles are usually dried back onto the grain, but there is considerable potential for inclusion of condensed solubles in liquid supplements due to the high phosphorus and nitrogen content. Many types of evaporation equipment can be used to remove water. Common types are multi-effect, mechanical recompression, short or long tubes, and forced or natural circulation.

The efficiency of the handling the whole stillage is centered around the removal of water. The ability to remove water is enhanced by the ability to separate the grain or fiber material from the thin stillage, which will raise the viscosity and in turn decrease the efficiency of evaporators. Evaporators offer the most efficient method for removal of water so a majority of the water removed should be done by the evaporators. Evaporators will be able to remove approximately 4 lb of water per pound of steam used.

The grain or fiber portion of the whole stillage (including the condensed solubles in some cases) are usually dried with rotary drum dryers, using steam tubes or direct fired. Grain dryers do not operate near as efficient as evaporators using approximately 1.2 to 1.8 lb of steam to remove 1 lb of water. This efficiency dictates the importance of efficient operation of centrifuges or grain separation.

Table 6
NUTRIENT COMPOSITION OF GRAIN AS
AFFECTED BY DISTILLATION (DRY BASIS)[7]

	Corn grain	Corn distillers	
		Dried grains	Dried solubles
Moisture (%)	11.0	8.0	10.0
Protein (%)	10.0	29.5	29.8
Fiber (%)	2.2	12.8	4.2
Fat (%)	3.5	8.0	—
Calcium (%)	0.02	0.10	0.30
Phosphorus (%)	0.32	0.95	1.60
TDN (%)	91	84	84
NE_{milk} (kcal/lb)	920	880	920
NE_{gain} (kcal/lb)	670	600	670

A cost usually not considered when planning an alcohol plant is the disposal of waste products from or resulting from processing of the whole stillage. These waste products can be the thin stillage in some instances or the liquid removed by the evaporators or the vapor vented to the atmosphere by the grain dryer. The vapor vented by the dryer may just have a foul odor and not require additional scrubbing. The liquids from the evaporators or centrifuge will be high in organic matter and possible suspended solids and most likely will have a high biological oxygen demand. Preventing their discharge into a natural watershed is critical. Disposal of this waste should be like any other waste treatment. Local and state regulations will most likely determine the proper method.

Table 6 shows the effect of fermentation and further processing using corn as an example. The removal of starch (accounting for two thirds of original material) results in the concentration of remaining nutrients in the recovered fiber and solubles (accounting for one third of original material). Livestock required to utilize the by-products of an alcohol plant can be determined from Table 7.

The value of distillers' dried grain or distillers' dried solubles will depend on a number of factors. The current market price for feed grain and market price for high-protein supplements such as soybean meal or urea will determine the market price range for distillers' grains. Distillers' grains are considered a commodity and many times prices can be found by contacting feed dealers and traders, elevators, and brokers.

IV. PLANT DESIGN

It is difficult within the limitations of this book to discuss all plant considerations when designing a plant. The basic requirements for producing alcohol have been covered. Every new location for an alcohol plant will provide different variables to contend with or allow for. In spite of this, there are questions that must be addressed and research undertaken before a project is put in motion.

What feedstock(s) will be used in the facility? When is it available? Where is it available? Will the plant require long-term storage? Large alcohol producers, like ADM, have the advantage of processing corn for a number of other products such as high-protein corn gluten and corn oil. They are also large enough to collect carbon dioxide. One product does not carry the full weight of the investment. A small plant will not have this luxury. The availability of a low-cost feedstock could be of benefit to the total economics of the plant.

What are the markets for products? This can only be reliably determined by contacting potential users or measuring internal demand. In projecting possible revenue from future

Table 7
ANIMAL POPULATION (IN THOUSANDS) REQUIRED TO UTILIZE THE DISTILLERS COPRODUCTS FROM VARIOUS SIZED ETHANOL PLANTS

Animal	Plant size (mm gal/yr)									
	5		15		25		50		100	
	Still-age[a]	DDGS	Still-age	DDGS	Still-age	DDGS	Still-age	DDGS	Still-age	DDGS
Calf (550 lb)	25	17	75	51	125	85	250	170	500	340
Steer (770 lb)	17	12	51	86	85	60	170	120	340	240
Cow (1,300 lb)	22	15	66	45	110	75	220	150	440	300
Pig (60 lb)	131	90	393	270	655	450	1,310	900	2,620	1,800
Pullets (3.7 weeks)	NA	758	NA	2,274	NA	3,790	NA	7,580	NA	15,160
Pullets (7.5 weeks)	NA	448	NA	1,344	NA	2,240	NA	4,480	NA	8,960

Note: Assumes corn feedstock and 330 days production per year.

[a] Ten percent solids.

From Solar Energy Research Institute, *A Guide to Commercial Scale Ethanol Production*, Washington, D.C., 1981.

sales, remember the fluctuations in the grain and fuel market which will set the price for both alcohol and distillers' dried grains.

Plant size and location will depend on decisions made on the first two questions. Location with easy access to feedstocks and delivery of products is critical. Availability of water and utilities and even labor should be considered. Remember that this is an industry capable of occasional foul odors and possibly will demand access to land for disposal of stillage.

Sizing of the plant and equipment will be first governed by the plant output. Operating schedules will next affect the size of equipment. Individual pieces will have to be sized according to the technology also.

Although very small-scale plants have been designed without professional assistance, it is not recommended that any project be undertaken without consulting individuals of firms that have experience in operations and plant design.

What type of energy feedstock will be utilized? In the long-term, tax benefits will not be available for alcohol plants which do not utilize renewable energy sources such as wood, coal, or solar. A number of biomass boilers are now on the market which make it feasible for small facilities to utilize sawdust, corn stover, or straw to generate steam. Remember that the cost of producing steam will affect the economic efficiency of almost every step in the production process. "Cheap Steam" can, in some cases, make up for energy-inefficient steps in the process.

Other questions of minor importance include availability and/or sophistication of labor supply? The legal or business framework in which you will finance, build, and operate the facility? Federal, state, and local permits required to operate?

V. CONCLUSIONS

This chapter should have brought to light the complexity of alcohol production. Technology has come quite a ways from making moonshine in old copper kettles. A modern alcohol plant just like a modern farm has to be run like a business. In the rush for so-called "energy independence", many American farmers attempted to build and operate a fuel alcohol plant. Few of these attempts got off the ground, and fewer yet are still in operation. Those still operating have had to enlarge in order to benefit from economics of scale. They have had to automate in order to increase efficiency of both operation and economics.

This is not intended to be an endorsement for only large facilities. The costs to construct and operate a small on-farm plant are prohibitive. This is not to say that conditions do not exist for a small plant (500,000 to 1,000,000 gal/year) which could utilize all the factors for efficient production and utilization of alcohol. Availability of large quantities of crop residue or sawdust, access to distressed or cheap grain, using the by-product from a food processing plant, or locating the plant in conjunction with a livestock operation would all help to make it more economically feasible.

Alcohol plants in the medium-size range (5 million to 20 million gal/year), although not usually funded by a small group, can be a big boost to surrounding agriculture and in turn the community at large.

Figure 11 shows an example of a small alcohol plant that took advantage of local conditions when considering the feasibility of a plant. It is located in the San Luis Valley of Colorado in a heavy agriculture area. Major local crops are barley, wheat, and potatoes. Barley that does not meet brewing standards and potatoes not suitable for packaging are the feedstocks for this plant, providing an alternate market for local growers. The timber industry in the nearby mountains offers tons of sawdust to power the biomass boiler, providing cheap steam.

FIGURE 11.   A 3 million gal/year plant in Monte Vista, Colo. Notice the sawdust pile in foreground. Waste sawdust from local lumber industry provides a cheap source of fuel. Cull barley and potatoes from area farmers and produce warehouses provide a source for feedstock.

REFERENCES

1. Information Resources, Inc., *U.S. Alcohol Fuels Data Base,* Washington, D. C., 1984.
2. **S.E.R.I.,** *Small Scale Fuel Alcohol Production,* Washington, D. C., 1980.
3. U.S. National Alcohol Fuels Commission, *Fuel Alcohol: An Energy Alternative for the 1980s,* Final Report, Washington, D. C., 1981.
4. **Fogarty, W. M. and Kelly, C. T.,** *Developments in Microbial Extracellular Enzymes,* 1979.
5. **S.E.R.I.,** *Fermentation Guide for Potatoes,* Washington, D. C., 1981.
6. **Perry, R. H. and Chilton, C. H.,** *Chemical Engineers' Handbook,* 5th ed., McGraw-Hill, New York, 1973.
7. **Poos, Mary I. and Klopfenstein, T.,** Nutritional Value of By-Products of Alcohol Production for Livestock Feeds, presented at the National Gasohol Commission Workshop in Columbus, Ohio, July 20—23, 1979.
8. **S.E.R.I.,** *A Guide to Commercial Scale Ethanol Production,* Washington, D. C., 1981.

Chapter 6

ANAEROBIC DIGESTION OF ANIMAL MANURE

J. R. Fischer, E. L. Iannotti, and J. Durand

TABLE OF CONTENTS

I. BACTERIOLOGY

Anaerobic digestion is a biological process in which microbes convert organic matter to methane and carbon dioxide. The process is usually depicted as occurring in four stages including metabolism of: (1) complex molecules to volatile fatty acids (VFAs), carbon dioxide, and hydrogen, (2) acids with three or more carbons to acetate, carbon dioxide, and hydrogen, (3) hydrogen and carbon dioxide to acetate; (4) carbon dioxide and hydrogen or acetate to methane.[12,72,136] The organisms that carry out the stages will be called acidogens, acetogens, homoacetogens, and methanogens, respectively.

This discussion of the microbiology and biochemistry of anaerobic digestion is considered an introduction, not an exhaustive review. For more complete reviews of anaerobic digestion of organic matter to carbon dioxide and methane see References 11, 12, 48, 50, 72, 73, 84, 89, 128, 136, and 141.

A. Bacterial Population

The total number of organisms in a normal digester is between 10^9 and 10^{10} bacteria per milliliter of digester fluid.[60,71,83,128] This concentration of bacteria is relatively high.[13,93,111,128] The total numbers of bacteria do not reach stable levels for a relatively long time after start up of a swine manure digester; in fact, the levels increase long after the biogas production becomes stable.[31] The final bacterial count in stable digesters is proportionate to the solids destroyed.[31,32,34,65]

The bacterial population in a digester is a complex mixture of many different species; however, only a few species are found in high concentration. The predominant organisms in a digester change with the type of substrate. The bacterial strains isolated from a swine manure digester are predominantly Gram-positive anaerobes; Gram-negative, non-spore-forming anaerobic rods predominate in domestic sludge and thermophilic cattle waste digesters.[61,71,83,119,120,128] In the case of manure digesters, the organisms are different from those found in manure.[61,109,128]

Although the populations differ from system to system, there are a number of similarities. For the most part the predominant organisms are strict anaerobes. Methanogens normally comprise 1 to 5% of the population in a stirred tank digester, while the percent of aerotolerant bacteria vary with the pretreatment of the waste.[31,32,50] A high concentration of Gram-positive organisms appears to be characteristic of swine manure, swine manure handling systems, and the swine manure digester.[61,109,114] An organism similar to *Bacteroides ruminicola* has been found in many anaerobic ecosystems; *Methanosarsina* has been observed in the swine manure and domestic waste digester.[57,82] The organisms best suited to carry out the process are present in a variety of systems, including the manure; however, they are not in high numbers. Thus, the digestion process is best started with material from an anaerobic process close to what is desired. The loading rate should be increased slowly to allow time for development of the final population.

B. Flow of Carbon

Anaerobiosis does not limit bacterial decomposition of cellulose, hemicellulose, pectin, and other polysaccharides but hinders metabolism of fatty acids, hydrocarbons, waxes, and aromatic compounds. Substances such as lignin are not degraded. Substrate components which contribute to methane production in a swine manure digester are, in decreasing order of importance, complex polysaccharides (hemicellulose and cellulose), lipids, protein, short chain acids, and starch.[65] Some 65% or more of these components in swine manure are converted to biogas, with the exception of protein, which exhibits about a 47% reduction during digestion. Dairy cattle wastes contain more cellulose and hemicellulose than swine manure; these components are degraded approximately 44%.[137] The final percentage of methane depends on the percentage of the original elements in the manure.

Lignin, silica, intrinsic characteristics of cellulose, essential oils, cutin, and polyphenols are related to the extent to which complex structures are degraded and determine the yield of biogas.[124] Of these, lignin, more than any other component, limits the ultimate extent of degradation. In the rumen, the indigestible portion of the feed is roughly 2.5 to 3.0 times the initial lignin content of the feed.[124] The predicted maximum destruction of swine manure based upon this value would be 85%.[65] The empirical destruction was found to be 83%.[57]

The acidogens initiate the breakdown of complex substances with exoenzymes. About 30% of the exoenzymes in the rumen are cell bound.[53] The most common generalization is that one part of the population (stenotrophs) initiates the degradation, then another part of the population (eurotrophs) aids in that degradation and proliferates on the polymer fragments produced by the stenotrophs.[11,54,138] The stenotrophs are specialists which are able to degrade the more resistant components and many times are limited to these substrates. While the eurotrophs have affinity for specific substrates, they are more versatile.[106] Complex polysaccharides are degraded to oligosaccharides and simple sugars; proteins are degraded to peptides and amino acids; lipids are degraded to fatty acids, glycerol, and other small molecular weight components. The polymer fragments then are transported into the bacterial cells where they are metabolized to short chain acids, carbon dioxide, and hydrogen. There are few free intermediates in digester fluid; the rate of utilization usually equals the rate of production.[57]

Methanogenesis from short chain acids is normally considered the rate-limiting step in the digestion of dissolved organics and the hydrolysis of insoluble polymers is rate limiting for the overall process.[12,37,39,70,96] The rate of digestion of insoluble polymers could therefore, theoretically, be increased to the maximum rate of conversion of the short chain acids to methane.

The rate of fermentation, unlike the extent of degradation, is not related to total amount of lignin but rather to intrinsic properties, arrangements in the cell, and associations with other molecules.[124] Chemically isolated cellulose is not uniformly digested. Thus, chemical analysis alone does not partition the fiber components into digestible and nondigestible fractions; the anaerobic process itself is required.

Organic matter in manure is for the most part insoluble. Swine manure contains only 7% soluble volatile solids;[57] 52% of the volatile solids are found in relatively small particles (<0.210 mm). Most of the cellulose and hemicellulose is found in larger particles (>0.210 mm). Lipids and proteins are found primarily in small particles. Methane yield is often reduced because the small particles are lost before addition to the digester and are difficult to retain in the digester.

Most of the large particles in swine manure and digester effluent are from feedstuffs fed the animal; these particles were comprised primarily of corn and soybean hulls.[57] While the surface of some particles contained a thick complex mat of bacteria, there was a significant number of surfaces on the larger particles to which organisms were not attaching and degrading. The surface was apparently composed of indigestible material and most of the digestive activity was occurring inside the particles. When digester contents were incubated for long periods of time, relatively large particles remained with resistant surfaces showing little evidence of microbial degradation. Thus, solids will build up even at long retention times.

An image of this microscopic environment can be gained from studies of forage breakdown in the rumen.[4,19,20] The bacteria first enter at the stomata, then spread through intercellular spaces. Entrance into the cells is gained by disruption of the cell wall. Easily degradable structures, mesophyl, and phloem are digested without attachment. Moderately resistant outer bundle sheaths and epidermal tissues are broken down more slowly. Highly lignified vascular and scherenchyma tissues are left intact and have very few adherent bacteria. Intracellular microcolonies with single or mixed morphological types are subsequently formed. These bacteria are in close contact with the plant material and specialize in what they degrade.

A gradation from easily degradable to more resistant compounds is also evident in manure as reflected by the rates of biogas production; there is a peak of activity in the digester immediately after loading.[57,81,94] The highest rates of biogas production occur within the first 4 hr after loading a digester. In the swine manure digester, 24% of the daily biogas is produced in the first 4 hr; 50% is produced within the first 9 hr of a 24-hr loading cycle.

For the most part, VFAs are not found in high concentration in the digester, with the acids being converted to methane as rapidly as they are formed.[21,57,81,94] The peak in the acid concentration occurs shortly after loading, with acetate being the most important intermediate in the flow of carbon to methane. Depending on the time after loading, 60 to 80% of the methane comes from acetate.[68,70,86,94,112] Propionate accounts for 20 to 40% of the methane produced; less methane is formed from butyrate. A small quantity of acetate (5 to 10%) is converted to carbon dioxide, presumably because a compound other than methane is a terminal electron acceptor. A similar amount of acetate is incorporated into bacterial cells.

Acetate, propionate, succinate, lactate, and ethanol are produced by pure cultures of acidogens from manure digesters.[61,128] As in the total population, acetate is produced in the greatest quantities and is the sole end product of many of the isolates.[61] However, interspecies transfer of hydrogen modifies the overall fermentation.[63,139] The interaction results in lesser quantities or an absence of reduced products such as ethanol, lactate, propionate, and succinate in the total population. If any succinate were produced, it would be rapidly metabolized to propionate; we have not found significant levels of succinate in any digester. The end result is that acetate, propionate, butyrate, and carbon dioxide turnover accounts for the major carbon flow in digesters. Propionate and butyrate are subsequently metabolized to acetate.[8,12,89]

Most methanogens utilize hydrogen and carbon dioxide while a few metabolize acetate.[6,15,82] The acetate-degrading system was saturated during the first 10% of the day in a swine manure digester.[64] This system in well-digested sludge was about half saturated.[70]

C. Nutritional Requirements of Bacteria

The digester bacteria have complex nutritional requirements. Digester bacteria require growth factors similar to those required by other anaerobic bacteria plus unknown factors in crude extract.[50,59,60,62,66,128] The most important nutrient besides the energy source is nitrogen. While both organic and inorganic nitrogen can be utilized, ammonia is usually an excellent source of nitrogen.[11,45,53,62,129,140] The ammonia usually results from the degradation of influent protein. A few anaerobes require preformed nitrogen sources. Bacteria like *Bacteroides ruminicola* require that amino acids be in the form of peptides with four or more amino acids.[103]

VFA's are important nutrients of anaerobic bacteria. For example, the cellulolytic ruminococci require branched-chain acids.[2,3,45] These acids come primarily from branched-chain amino acids and are used to synthesize protein and longer branched-chain fatty acids and aldehydes. Hungate and Stack[55] reported that selected strains of *Ruminococcus albus* also require phenylpropionate as a growth factor.

Bacteroides are known for their requirement of hemin, vitamin K, and menadione.[17,78] These compounds result primarily from degradation of plant material and are important as a component of cytochromes. *Succinivibrio dextrinosolvens* requires naphthoquinone.[40] Most anaerobic bacteria require vitamins, especially B vitamins.[11,53] Requirements change with the substrate and carbon source.[79]

Methanogens have relatively simple growth requirements in that few if any organic substances are needed.[14,84,136] Many can grow with carbon dioxide, hydrogen, ammonia, and trace metals. Acetate and coenzyme M are notable required compounds.

Very little is known about the mineral, especially trace mineral, requirements of anaerobes.

Hydrogen sulfide is the most important inorganic source of sulfur.[11,45,94,98,129] Cysteine is the most important organic source. Inorganic forms of phosphorus can be used over a wide range of pH. *B. ruminicola, B. succinogenes, and B. amylophilus* and several other genera have an absolute requirement for sodium.[16] Some bacteroides have an absolute requirement for potassium. All methanogens tested require nickel, which is taken up by an active transport system.[67]

Many rumen bacteria require low levels of carbon dioxide which is used as a cellular constituent.[11,53] Succinate producers use large amounts of carbon dioxide. A cellulolytic sludge bacterium was inhibited by 100% hydrogen or 20% carbon dioxide in hydrogen.[108] Of course, the methanogens require carbon dioxide and hydrogen.

Understanding of the bacteriology of anaerobic digestion is minimal compared to our knowledge of the rumen. For example, many bacteria that carry out the anaerobic process have not been isolated. Typically, 15% or less of the total number of bacteria in beef cattle and domestic digesters as determined by microscopic count can be enumerated and cultured.[60,71,83,128] While 60% of the organisms in a swine waste digester have been enumerated and isolated,[60] 40% of the organisms present cannot be isolated. Also, many of the organisms grow only to a limited extent once isolated. The need for crude extracts in the media reflects a lack of knowledge and a reduced ability to define the optimum conditions for methane generation. The inability to grow many of the organisms may involve not only a failure to supply the proper nutrients but also to duplicate the interactions that take place in the total population.

In the rumen, bacteria associated with the organisms that initiate fiber breakdown supply growth factors to these specialists.[138] Our results indicate a similar relationship in the digester. The spent media from cultures of our largest group of strains promotes growth of strains associated with fiber degradation. There are also interactions involving methanogens.[7] Some interactions are more than nutritional exchanges; they include actual physical contact.[56] *Methanosarcina* are closely associated with other morphotypes. In the future we will probably show that the longer retention and controlled growth in digesters has permitted development of complex requirements; the activity in the digester is the result of teams with more interactions than other anaerobic ecosystems.

D. Failure of Digesters

The reasons for process failure are (1) toxic compounds produced by the microbial population or (2) added toxic compounds. The former commonly results from an imbalance in the microbial population and results in products such as ammonia and VFAs. The buildup of these microbial products, especially VFAs, is an accepted indicator of instability. While the cause is usually an outside perturbation, the end products themselves can be toxic and lead to further deterioration of the process. Also, there is a complex interaction between the end products and the physical environment. The interaction involves pH, alkalinity, the carbon dioxide pressure, VFAs and ammonia, and other ionic substances present.

The pH in a digester should remain near neutrality.[85] A drop in pH is usually more serious than an increase. In a swine manure digester the pH remains near 7.4 and decreases below 7 only in severely stressed digesters. This is above the lower limits of the organisms even though the lower limit of pH of both the acidogens and methanogens from digesters are relatively high (pH 6).[57,61,108,122] Bacteria from other anaerobic ecosystems have a wider range of pH than that reported for digesters.[107] The cellulolytic and amylolytic bacteria are more sensitive to reduced pH than the other organisms.

The concentration of total VFAs is normally below 300 mg/ℓ, while stressed digesters contain 4000 to 5000 and up to 14,000 mg/ℓ.[26,32,36,75,86,88] A good rule of thumb is that the total volatile acid concentration should be below 500 mg/ℓ.[36,87]

VFAs are not equal in their effect on digester bacteria. Acetate is the least toxic of the individual volatile acids for acidogens from a swine manure digester.[57] Propionate and butyrate are more toxic but affect each group differently. Methanogens are less affected by acetate than propionate.[51,122] There are also indications that propionate is most toxic to the total digester population, with the affected organisms being the acidogens.[5,86] The unionized portion of the VFAs is the toxic factor.[75,110]

Bryant[12] has predicted that increases in the level of hydrogen would result in increases in the concentration of propionate followed by increases in butyrate. Propionate has been associated with instability more than other acids; however, increases in hydrogen have not always been associated with increases in propionate. This probably reflects the variety of events leading to digester failure.

High levels of ammonia are toxic, although only very high levels totally stopped growth of pure cultures of acidogens and methanogens.[51,57] Low levels of ammonia are required for growth. In sewage digesters, ammonia nitrogen concentrations become inhibiting between 1500 to 3000 mg/ℓ depending upon the pH; this corresponds to a free ammonia level of 150 mg/ℓ.[85,88] Hein et al.[44] reported problems in their pilot-size digester with beef manure when ammonia nitrogen concentrations exceeded 2000 mg/ℓ. Ammonia nitrogen concentrations of 2000 to 3000 mg/ℓ are normal in swine manure digesters.[32,110,125] Gas production was reduced but continued in a swine manure digester with up to 7000 mg/ℓ.[36,74,127] We also observed a reduction in volatile solids destruction at this level.[142] Poultry digesters operate at even higher levels, with up to 10,500 mg of ammonia nitrogen/ℓ being reported.[105]

Tolerance to potentially toxic cations such as ammonia is produced by acclimation of the microorganisms to the agent and by antagonism of other cations in a multiple-cation system.[76] Ammonia actually plays a dual role in digesters.[36] Besides being toxic, ammonia buffers volatile acids by maintaining a high level of bicarbonate. Increasing the level of carbohydrate decreases the percentage of nitrogen present as ammonia. Thus, the level of a toxic compound can be reduced but at the same time the bicarbonate buffer and the stability of the process is reduced.

Antibiotics, heavy metals, and other toxic compounds can poison the anaerobic process.[9,58,104,130,131] Turacliff and Cuter[121] and Fisher et al.[26] reported severe problems with theraputic levels of lincomycin in a swine manure digester. Most of the time the biogas production drops temporarily following a change in growth-promoting levels of antibiotics. The various antibiotics are not equal in their effect. Products such as paint removers can also inhibit methane formation.[21,116,118]

We have found that sudden changes in temperature, especially increases, are detrimental to the anaerobic process; however, given enough time the total population will adapt to the temperature of the digester. Thus, it is not surprising to find that the optimum temperature of pure laboratory cultures of digester bacteria is close to that of the system from which they were isolated.[108,122] Methanogens can be found in ecosystems at wide ranges of temperature.[141,122]

E. Mathematically Modeling the Digestion Process

Most of the models for anaerobic fermentation begin with a mass balance relationship such as: Rate of accumulation = Rate of material in − Rate of material out ± Rate of reaction. Many of these use the Monod[92] equation for bacterial growth. Anaerobic fermentation is a complex process; therefore, choosing the state variables about which to write a mass balance is difficult. Since the mass balance equations are simultaneous nonlinear ordinary differential equations and generally not soluble, steady state solutions or computer simulations have been used. The steady state solution is sufficient for use as a design tool.[43]

Defining the digestion process as quasi-steady state requires use of the dynamic modeling equations. The resulting set of simultaneous algebraic equations lend themselves to the graphical depiction of intervariable dependencies. Both steady state and dynamic models can predict steady state digester outputs (like gas production) with accuracies within $\pm 10\%$ and often within $\pm 5\%$. The steady state models make no attempt to explain dynamic effects of the digestion process, but even the dynamic models are at best semiquantitative in their description of dynamic digester behavior.

Digester modeling may be divided into four components: substrate characterization, bacterial characterization, inhibition, and output variables. Although several methods of characterizing the influent waste or substrate exist, most researchers have chosen volatile solids (VS). Hashimoto[43] has specifically used biodegradable volatile solids (BVS) in his model. In Hill's simplified model,[46] BVS and VFA content were used. Finally, in Hill's more complex model,[47] a number of quantities were required including BVS, VFA concentrations, NH_4 concentration, and others. In general, any time a mass balance is written for a quantity other than a bacterial group, an influent value of this quantity must be known. So the more complex the model, the more complex the waste characterization needed. As an example, Hashimoto's model has only one substrate mass balance given as:

$$\frac{dS}{dt} = \frac{S_0 - S}{\Theta} - \frac{\mu X}{Y}$$

where S = substrate concentration in the digester, S_0 = influent substrate concentration, Θ = hydraulic retention time, μ = specific growth rate, X = concentration of bacteria in the digester, and Y = growth yield coefficient. Therefore, S_0 completely characterizes the waste. A slightly more complex model by Hill[46] has two substrate mass balances given as:

$$\frac{dS}{dt} = \left(\frac{S_0 - S}{\Theta}\right) - \frac{\mu M}{Y}$$

$$\frac{dAC}{dt} = \left(\frac{AC_0 - AC}{0}\right) + \frac{\mu M}{Y}(1 - Y) - \frac{\mu_c M_c}{Y_c}$$

where M = concentration of acid-forming bacteria, Y = growth yield for acid-forming bacteria, S = concentration of BVS in the digester, S_0 = influent concentration of BVS, μ = specific growth rate for acid farmers, AC = VFA content in the digester, and AC_0 = influent VFA. The c subscript refers to methanogenic quantities. In this model, two quantities S_0 and AC_0 must be given to specify the waste.

As presented earlier in the chapter, several bacterial groups are involved in the digestion of organic wastes; however, most models include only a limited number of mass balance equations for bacteria.[35,43,47,97]

Hashimoto[43] described inhibition based on influent substrate concentrations (S_0). At influent solids concentrations above 80g VS ℓ^{-1}, a kinetic constant exponentially increases as a function of S_0. VFAs and ammonia concentrations were modeled by Andrews[5] and Hill.[47] Although not inhibitory, death of bacteria can cause a decrease in the yield coefficient, especially at low concentrations of bacteria (as when hydraulic retention time decreases).

Steady state models, such as Hashimoto's can predict the steady state values of substrate concentration and bacterial mass. More detailed models, such as Hill's dynamic model,[47] can predict over ten parameters as functions of time, including concentrations of four different microbial populations and the concentration of VFAs.

At this time only limited comparisons have been made among the various models. All investigators claim good agreement with selected steady state data, but many times prediction of dynamic digester behavior has not been satisfactorily demonstrated. This is an indication that the anaerobic fermentation process is not completely understood.

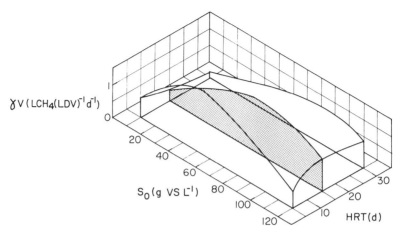

FIGURE 1.   The relationship between liters of methane per liter of digester volume per day and HRT vs. influent VS concentration added to the digester. The shaded area represents the HRT at which data in this experiment was collected. (From Fischer, J. R., Iannotti, E. L., and Porter, J. H., *Agric. Wastes,* 11, 157, 1984. With permission.)

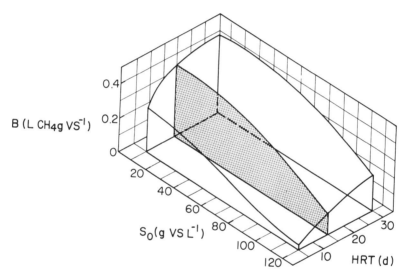

FIGURE 2.   The relationship between liters of methane per liter of digester volume per day and HRT vs. influent VS concentration added to the digester. The shaded area represents the HRT at which data in this experiment was collected. (From Fischer, J. R., Iannotti, E. L., and Porter, J. H., *Agric. Wastes,* 11, 157, 1984. With permission.)

II. DIGESTERS

A. Design of Methane Plants

Biological as well as engineering parameters must be considered in the design of methane plants. The design must account for the physical, biological, and chemical makeup of the substrate to be converted into methane, and the mass, fluid, and energy flow of the system. The optimization of the design for high methane production rates (liters of methane per liter of digester) or high methane yields (liters of methane per kilograms VS added to the digester) must be selected since these criteria are inversely related. High methane production rates can be obtained from high organic loading rates and very short hydraulic retention times (HRT), whereas high methane yields can be obtained from low solid concentrations and long microorganism retention times (Figures 1 and 2).[30]

Table 1
TYPES OF DIGESTER CONFIGURATIONS

Retention characteristics	Reactor examples
Liquid = solids = microorganisms	Continuously stirred tank, plug flow and batch fed
Microorganisms and solids is greater than the liquids	Continuously stirred tank with solids recycle, upflow solids, upflow sludge blanket, and baffle flow
Microorganisms greater than the solids in the liquid	Anaerobic filter and fluidized and expanded bed

B. Digester Configurations

To achieve different HRT, solids retention times, (SRT), and microorganism retention times, (MRT), various physical configurations of digesters can be built (Table 1).[25]

The majority of agricultural digesters are intermittently mixed vertical tanks or horizontal plug flow designs. In these designs the retention time of the microorganisms, the solids, and the liquid tend to be the same. Gilman et al.[38] reported that in northern New England both vertical tank and horizontal plug flow digesters are being installed on dairy farms. Cournoyer et al.[24] modified the plug flow digester design by incorporating a mixing system consisting of biogas injected at the bottom of the digester to alleviate crusting problems. Pigg and Vetter[101] reported on a vertical tank intermittently mixed digester that has operated for several years on a commercial dairy farm in Minnesota. The potential gross energy production from this unit was 3 kWh per cow per day. Some recent work by Hasheider and Sievers[42] and Brum and Nye[10] tested an anaerobic filter for digesting dilute flush wastewater from animal confinement buildings, but these systems have not been put into production facilities. In the batch fed digester, the total feed volume is added to the digester in a single load and remains there until the digestion is completed. This is a noncontinuous process and is best suited for seasonally produced material.

Methods of agitating vertical tank digesters usually consist of mechanical stirrers or gas recycling with some other limited options. Advantages of a continuously stirred tank reactor are (1) it processes manure with high suspended solids, (2) it increases the contact of the substrate with the microorganisms, (3) it establishes even temperature distribution throughout the tank, and (4) it prevents scum layer formation. The disadvantages of this type of reactor are (1) power requirements to operate the mixing mechanism, (2) the possibility of undigested substrate leaving the digester, and (3) microorganisms leaving the digester.

Ideally, in plug flow digesters the material passes through the digester in a plug-like manner with influent entering at one end and effluent removal at the other. The only mixing that occurs in these digesters is the vertical mixing due to gas production. Some settling of the solids can also take place in these digesters, resulting in longer SRTs than HRTs. The major advantages of a plug flow digester are (1) easily installed on an operating farm, (2) fairly high conversion, and (3) high stability. The disadvantages are (1) the possibility of the digester plugging with solids, (2) difficulty maintaining temperatures, and (3) possibility of formation of a thick crust. One way to prevent the solids from settling is to use baffles in the digester, thereby forcing the solids out of the digester with the liquids. In one study of a plug flow digester operated at ambient temperature, gas production was one third that of a conventional digester. However, the net gas yield is comparable since there is no energy requirement for heating.[134]

Since the methanogens grow at a different rate and have different optimum conditions than the nonmethanogenic bacteria, it is possible to design two digesters with optimum characteristics for each of these groups. End products of the acidogen (VFAs) can then be fed into the methanogen digester in a continuously controlled manner. One study reported a significantly higher methane to carbon dioxide ratio when the methane phase was isolated than when both phases occurred in the same digester.[80] The two-phase digestion system has the ability to produce and enrich methane product gas and it has a greater stability with respect to feed stock, loading, pH, and shock. Also, it has the ability to maintain appropriate densities of acidogens and methanogens. The main disadvantage of the two-phase system is the cost. Capital expenditures for additional equipment, controls, and higher labor requirements make it less applicable to farm size operations.

C. Designing the Digester System

In order to adequately design a digester for an agricultural operation the design engineer must select: (1) process design and structure, (2) loading rate, (3) HRT, and (4) temperature. Also, the equipment and facilities for integrating the system into the agricultural production enterprise and the energy consumption patterns of the farm must be considered.

D. Process Design and Structure

As previously mentioned, most agricultural digesters are incorporated into livestock enterprises. The continuous digestion process adapts well to livestock systems, since the bacteria in the digester must be fed daily and livestock produce manure continuously. The advantages of an intermittent vertical tank reactor vs. horizontal plug flow reactor have not been clearly defined. Thus, selection of one over the other is mainly site specific.

The type of material used for the digestion container can be quite variable, with the major concerns being availability, gas and water tightness, and the necessary strength to withstand the water and gas pressures. Most farm digesters have a fixed roof[31,133] with the exception being the flexible membrane-covered horizontal plug flow digester.[24,69] Insulation has been successfully used both inside and outside of the digester when adequate protection for the insulation was provided.[31,102]

Most piping material has been PVC, except for heat exchanger use. For efficient fluid flow, pipe dimensions should be 76 mm, although Goodrich[41] successfully used 38-mm piping with swine waste. Fischer[31] and Pearson[99] both reported that 90° elbows in the piping carrying manure slurries caused plugging.

In selecting pumps for transferring manure solids into the digester, submersible pumps with choppers[31,101] and ram pumps[99] have been used most successfully. Common problems encountered in pumping the slurries have been freezing, foreign material such as metal in the slurry, and excessive bedding, especially with free stall dairy waste.

E. Loading Rate

The quantity of manure added to the digester divided by the liquid volume of the digester is the loading rate.

$$\text{Loading rate (LR)} \left(\frac{\text{kg VS}}{\text{m}^3}\right) = \frac{\text{kg VS added to digester/day}}{\text{Volume of digester (m}^3)}$$

The quantity of VS produced by various animal species is presented in Table 2. Generally, the LR is dependent upon the animal species from which the manure was collected, the ration fed the animals, and the age of the manure. Chicken manure is most digestible, followed by swine manure, with beef and dairy being least digestible. General guidelines for loading rates from published reports are presented in Table 3.

F. Retention Time

The HRT is the number of days for replacement of the fluid volume of the digester.

Table 2
DAILY MANURE PRODUCTION FROM VARIOUS
ANIMAL SPECIES[a]

Animal	Size (kg)	Total solids (kg)	Volatile solids (kg)
Dairy cattle	68	0.73	0.59
	113	1.2	0.95
	227	2.4	1.9
	454	4.7	3.9
	635	6.6	5.4
Beef cattle	227	1.6	1.4
	340	2.4	2.0
	454	3.2	2.7
	567	3.9	3.4
Cow		3.3	2.8
Swine			
Nursing pig	16	0.09	0.08
Growing pig	29	0.18	0.14
Finishing pig	68	0.41	0.33
	91	0.54	0.44
Gestate sow	125	0.37	0.30
Sow and litter	170	1.4	1.1
Boar	159	0.45	0.38
Sheep	45.4	0.45	0.39
Poultry			
Layers	1.8	0.024	0.017
Broilers	0.9	0.016	0.011

[a] Data obtained from Midwest Plan Service, *Livestock Waste Facilities Handbook,* MWPS-18, Department of Agricultural Engineering, Columbia, Mo.

Table 3
VARIOUS LOADING RATES FOR
INDIVIDUAL ANIMAL SPECIES

Animal species	Loading rates (kg volatile solids m^{-3})
Swine	4.8
Beef	5.6
Dairy	7.2
Poultry	2.1

Note: These are average loading rate values that have been obtained from literature.

$$\text{Hydraulic retention time (days)} = \frac{\text{Digester volume (m}^3)}{\text{Volume of influent (m}^3/\text{day)}}$$

HRTs for various animal manures have been shown to be in the range of 15 to 30 days. As previously discussed, after approximately 25 days the majority of the materials has been degraded.

In a continuously stirred tank reactor the HRT is the same as the SRT. At SRTs of up to 25 days the degradation of volatile solids follows the first order decay model.[43] However, at (SRTs) greater than 25 days, most manure substrates have been degraded.

At longer retention times the nutrient requirements for the bacteria are less than those for shorter retention times; this could be due to the increased bacterial cell synthesis and therefore the increased degradation of nutrient-containing compounds. Some digester designs, such as the attached film, increase bacterial retention time, thereby preventing washout of the slow-growing methogenic bacteria and reducing the nutrient requirements for cell synthesis.

The relationship between the concentration of solids in the influent, the loading rate, and the HRT is given by the following equation:

$$VS = (LR)(HRT)$$

where VS = volatile solids in the influent, LR = loading rate as grams volatile solids per liter, and HRT = hydraulic retention time in days.

A maximum solids concentration of 32% has been determined for the methanogens whereas for the acid-producing bacteria the maximum concentration has been 60%.[1]

Two objectives must be considered in selecting a LR and a HRT — the optimum for energy production and the optimum for odor reduction. The longer the HRT and the lower the LR, the less odorous is the effluent.[125,135] For optimum energy production, two measures of efficiency to consider are percentage of VS destroyed and the quantity of gas produced per unit volume of digester. These two measures of efficiency are inversely related.[34] As gas production (cubic meter of gas per cubic meter of digester) increases with increasing LR, the solids destruction efficiency (cubic meter of gas per kilogram VS added) decreases. For swine manure digested at a temperature of 35°C and 4 kg VS m^{-3} day^{-1} LR, 0.56 m^3 of gas was produced per kilogram VS added to the digester.[32] Van Velsen[125,126] reported that a LR of 3.6 kg VS m^{-3} day^{-3} and a 15-day HRT were optimum with respect to both gas production and the reduction of malodorous compounds (temperature 35°C).

G. Temperature

The quantity and quality of biogas produced are directly affected by the temperature of the digesting slurry. A temperature range of 30 to 35°C appears to be optimum for anaerobic digestion of swine manure. Digesters have been operated at both mesophyllic (25 to 40°C) and thermophyllic (40 to 65°C) temperatures;[32,77,91,126] however, most of the studies have been done in the mesophyllic range. VanVelsen[126] reported 25% less gas with thermophyllic digestion of swine manure than with mesophyllic digestion; gas production increased linearly with temperatures from 25 to 40°C. The optimum temperature in this range would depend on the net energy efficiency of the system, including conduction losses and thermal energy required to heat the influent. The van Velsen[126] studies were performed with manure from a barley ration; the effect of diet on operational parameters is not known.

H. Biogas Produced

The amount of biogas produced from the animal manure digesters depends upon several factors:

- Type of animal from which the manure is collected
- Quality of the manure
- The addition of bedding with the manure
- The ration fed to the animals
- The manure handling system
- The layout of the farm
- The design criteria of the digester

Table 4 gives a range of biogas production for different types of manure.

Table 4
BIOGAS PRODUCTION FROM
VARIOUS TYPES OF MANURE

Manure type	Biogas production (m³/kg volatile solids added)
Beef — dirt lot	0.25—0.30
Beef — barn	0.37—0.44
Dairy — stanchion	0.50—0.56
Dairy — free stall	0.30—0.37
Poultry	0.37—0.56
Swine	0.37—0.56

I. Startup Procedure

For efficient digestion of an organic substrate, a stable, synergistic microbial population must be developed. During startup, the LR should be increased in increments;[27,49,77,127] we recommend starting at a loading rate of 2.5 g VS ℓ^{-1}day^{-1} and increasing the rate in weekly increments of 0.8 g VS ℓ^{-1}day^{-1}, until the desired loading rate of 4 g VS ℓ^{-1}day^{-1} is reached. Lapp et al.[77] reported a minimum of 18 days for starting swine manure digesters that were seeded with municipal digester effluent. Van Velsen[127] indicated that the time for stabilization could be as long as 180 days. Fischer and Iannotti[33] have indicated that an average time for steady state production of biogas is about 35 days, although establishment of a stable microbial population may take several months.

J. Management

Manure should be added to the digester at least daily to minimize the loss of potential methane due to degradation prior to addition to the digester. The most applicable methods of removing and transporting manure from livestock buildings are hydraulic flushing and mechanical scrapers. The manure/flush water mixture leaving a confinement building with a flushing system is 1 to 2% solids. A procedure that will increase the total solids concentration to approximately 8% is necessary for optimal loading of the digester. A concentration of 7.5% VS corresponds to a loading rate of 4 g VS ℓ^{-1} and a rentention time of 15 days, which are optimum for a swine manure digester. An economical method of obtaining the desired solids concentration is sedimentation; however, Sievers et al.[110d] have determined that as much as 30% of the degradable VS can be lost in the supernatant as the solids content is increased from 1.5 to 8% in a sedimentation process. Smith[113] calculated that the slurry removed by scraping is approximately 8% total solids, and hence suitable for introduction directly into the digester.

The most economical, convenient, and reliable method for determining digester performance is to measure gas production. If continual decreases in gas production are observed for several days, the physical system should be examined for leaks, drops in temperature, decreases in loading rates, or an introduction of toxic compounds in the digester. A range of chemical parameters associated with upset and failing swine manure digesters have been determined.[64] However, no clear patterns have been found that will provide forewarning and indicate the specific causes of the problem.

The daily volume of biogas produced has usually decreased with changes in composition of the feed or feed additives. On two occasions, severe decreases in the rate of biogas production were associated with the presence of an antibiotic.[26] Until more is known about the inhibitory concentrations and the ability of the population to adapt, the digester should be by-passed during the time that animals are receiving therapeutic levels of antibiotics.

Additionally, feed additives should not be changed when the system is already stressed by other factors. Fischer et al.[33] reported that once the decrease in biogas has become severe, stopping the loading for several days, diluting the volume of the digester 50 to 75% with water, then loading at a reduced rate has resulted in system recovery. They emphasize problems have been minimal, however, and the systems have been very stable.

K. Potential Uses of the Biogas and Effluent

The biogas produced normally consists of 60% methane and 40% carbon dioxide, with traces of hydrogen sulfide and other substances. The use of biogas as an alternative fuel is hampered by its physical properties. It has a low energy density of 16 kJ/m³ at atmospheric pressure and liquefaction occurs at 47 bars of pressure and $-82.5°C$. Therefore, liquefaction is not feasible in a rural setting and storage at atmospheric pressure is costly due to the large storage volume required per unit energy. Compression reduces the storage volume per unit of energy, but is expensive. However, 10 m³ of gas compressed to 200 bars (3000 psia) has a mass of 65 kg (140 lb) and contains an equivalent 6.2 ℓ (1.64 gal) of diesel fuel.

The most economical design for gas utilization is to simply use the gas as it is produced. One way to accomplish this is to use the gas for space heating needs on the farm. However, since space heating needs are seasonal, this method of direct utilization is not ideal. The use of the gas as a fuel in an internal combustion engine coupled to an electrical generator is perhaps the best method of direct utilization. The engine generator can convert the gas into electrical energy and the excess thermal energy from the engine coolant water and exhaust system can be collected and converted into thermal energy for use in the livestock buildings. Some rules of thumb for the number of animals required to produce 1 kw of electricity are 10/1400 lb dairy cows, or 2000/4 lb layers, or 160/120 lb finishing hogs, or 28/700 lb beef finishing steers. Some innovative designs of integrating a digester with an internal combustion engine coupled to an electrical generator with the excess heat being used for liquid fuel production have been shown to be technically feasible.[32]

A model swine farm will be described in the following discussion to illustrate the operating parameters and discuss the economics of anaerobic digestion. We will assume the model farm was designed such that the energy demand of the swine buildings matches the capabilities of the manure digester.[29] The selected pork production system markets 3200 hogs per year and has 250 sows. The farrowing and nursery buildings are environmentally controlled, and the gestation and finishing units are open front, naturally ventilated buildings. The manure is removed from the buildings to transport trenches by mechanical scrapers (one 373-W electric motor and two scrapers in each building). A drag chain transports the manure down the trench to the collection basin, from which it is pumped into the digester. Digested manure is stored in an effluent storage container until weather permits field application.

The digester is a cylinder 4.3 m in diameter and 6.7 m in height. The volume of the digester (96.6 m³) is calculated based on a loading rate of 4 kg m^{-3}day^{-1} VS plus an additional 10% gas volume above the liquid contents. The 351 kg/day of manure VS was estimated by assuming 0.1 kg of manure VS per kilogram of feed and a requirement of 4.0 kg of feed per kilogram of pork for the 3200 hogs marketed at 100 kg.[52,113]

Each kg of VS added to the digester yields 0.56 m³ of biogas (15 day retention time, 35° C, 4 g VS m^{-3}day^{-1}, 59% methane).[32] If the yield is reduced 12.5% to account for inefficient solids removal and gas production losses, the 351 kg of manure VS would give 197 m³ of biogas or 102 m³ of methane per day. Since the density of methane under standard conditions is 71.4 kg/m³ and the lower heating value is 50.2 MJ/kg,[115] the energy produced by the digester is 3.653 MJ, or 42.3 kW.

The methane is converted to electrical and thermal energy by a 12.5-kW Kohler electric generator connected to a Waukesha spark-ignited engine, model VRG 155U. The thermal energy was calculated assuming full recovery of heat in the jacket water and 50% recovery

Table 5
VARIOUS CAPITAL COSTS AND ENERGY PRICE OPTIONS INVESTIGATED[a]

Electricity ($/kWh)	Propane		Capital invested ($)	Annual return ($)	Net present value ($)	Benefit/ cost ratio	Breakeven capital investment ($)	Breakeven return ($)
	($/ℓ)	($/gal)						
0.04	0.15	0.55	42,250	1,962	− 12,287	0.71	21,380	4,172
0.04	0.15	0.55	62,375	1,962	− 23,455	0.62	21,380	6,181
0.06	0.22	0.85	42,250	4,317	807	1.02	43,057	4,172
0.06	0.22	0.85	62,375	4,317	− 10,362	0.83	43,057	6,181
0.08	0.29	1.10	42,250	6,552	13,233	1.31	64,297	4,172
0.08	0.29	1.10	62,375	6,552	2,065	1.03	64,297	6,181
0.08[b]	0.29	1.10	62,375	7,428	6,935	1.11	72,900	6,181
0.08[c]	0.29	1.10	62,375	5,676	− 2,806	0.96	58,394	6,181

[a]   Maintenance costs were at $0.30 per hour, except where noted.
[b]   Maintenance costs at $0.20 per operating hour.
[c]   Maintenance costs at $0.40 per operating hour.

of that in the exhaust gases. The engine-generator supplies all of the electrical energy and 93% of the thermal energy required on the farm.

In the economic analysis, the cost of the buildings was not considered since the buildings were not modified. The capital cost of the digester, engine-generator, and associated equipment was $62,375. Additional costs included oil for the engine, a yearly overhaul of the engine, and overhaul of the generator every 2 years, and enough propane to make up the 7% deficit in thermal energy. These additional costs were $3005 per year.[28]

The returns from the system included the energy and tax credits. The generator produces 6.8 kW of electricity; this value did not include electricity required for the digester. Only the thermal energy supplied at times of need was credited to the digester. The equivalent quantity of propane required for 11 kW of heating for 273 days in the farrowing house and 3.4 kW of heating for 121 days in the nursery was 16,440 units. The efficiency of a hot water boiler was assumed to be 70%.

Some 20% of the investment cost qualified as a tax credit (10% investment and 10% energy tax credit). For the economic analysis, the pork producer was assumed to be in the 30% tax bracket and desired a 10% rate of return in his investment. Straight-line depreciation techniques were used for the investment analysis. A 70% salvage value on the engine and generator and 10% on the digester were selected. The life of the system was assumed to be 12 years.

The analysis of the $62,375 investment in the digester-engine-generator system was proved to be profitable at costs of $0.29 for propane and $0.08/kWh for electricity (benefit to cost ratio of 1.03).[28] The break even capital requirement for the investment could increase to $64,297, or the benefits could decrease to $6181.

Other price scenarios are given in Table 5. The data in Table 5 indicate that profitability for the energy system at estimated current prices of $0.04 per kilowatt-hour for electricity and $0.15 per kilowatt hour for propane would require a reduction in investment to $21,380. If electricity costs should increase to $0.06 per kilowatt-hour and propane to $0.22, then $42,250 would be a profitable investment. The last two lines of Table 5 indicate the effect of varying the annual maintenance cost from $0.20 to $0.40 per operating hour. The benefit to cost ratio indicates that profitability can be achieved at $0.20 per operating hour (B:C = 1.11), but is not profitable at $0.40 per operating hour (B:C = 0.96).

The effluent from the digester contains 40% of the total solids that were in the manure entering the digester. A possible use of these residual solids is for animal feed. With swine manure digesters, the amino acid content of the effluent is slightly better-quality feed than that of the manure put into the digester. However, the amount of organic nitrogen remaining in the effluent is only 53% of the organic nitrogen in the influent.[65]

Utilizing the effluent as a source of plant nutrients is feasible; however, the concentration of these nutrients is low and thus expensive to incorporate into the soil. Fischer et al.[34] have shown corn yields from effluent fertilizer plots to be equal to commercial fertilizer and swine manure from a pit when applied on an equal nutrient basis.

Gilman and Bennett (1984) reported that economics of digesters on dairy farms is not fully dependent on fuel gas production. The value of the solids separated from the effluent of the digester were valued at $17,000 per year when used for bedding.

REFERENCES

1. **Abeles, T. Ellsworth, D., and Genereaux, J.,** Task VI, Biological Production of Gas for Office of Technology Assessment, Washington, D. C., April 1979.
2. **Allison, M. J., Bryant, M. P., and Doetsch, R. N.,** Conversion of isovalerate to leucine by *Ruminococcus flavefaciens, Arch. Biochem. Biophys.,* 84, 246, 1959.
3. **Allison, M. J., Bryant, M. P., Katz, I., and Keeney, M.,** Studies on the metabolic function of branched-chain volatile fatty acids. Biosynthesis of higher branched-chained fatty acids and aldehydes, *J. Bacteriol.,* 83, 1084, 1962.
4. **Akin, D. E.,** Microscopic evaluation of forage digestion by rumen microorganisms — a review, *J. Anim. Sci.,* 48, 701, 1979.
5. **Andrews, J. F.,** A. dynamic model of the anaerobic digestion process, Proceedings 23rd Industrial Waste Conference, Purdue University, Lafayette, Ind., May 1968.
6. **Balch, W. E., Fox, G. E., Magun, L., Woese, C. R., and Wolfe, R. S.,** Methanogens; reevaluation of a unique biological group, *Microbiol. Rev.,* 43, 260, 1979.
7. **Baresi, L., Mah, R. A., Ward, D. M., and Kaplan, I. R.,** Methanogenesis from acetate; enrichment studies, *Appl. Environ. Microbiol.,* 36, 186, 1978.
8. **Boone, D. R. and Bryant, M. P.,** Propionate-degrading bacterium, *Syntrophobacter wolinii* sp. nov. gen. nov., from methanogenic ecosystems, *Appl. Environ. Microbiol.,* 40, 626, 1980.
9. **Brumm, M. C., Sutton, A. L., and Jones, D. D.,** Effect of dietary arsonic acids on performance characteristics of swine waste anaerobic digesters, *J. Anim. Sci.,* 51(3), 544, 1980.
10. **Brum, T. J. and Nye, J. C.,** Dilute swine waste treatment in an anaerobic filter, presented at the 36th Purdue Industrial Waste Conference, Purdue University, Lafayette, Ind., 1981.
11. **Bryant, M. P.,** Microbiology of the rumen, in *Duke's Physiology of Domestic Animals,* 9th ed., Sevenson, M. J., Ed., Cornell University Press, Ithaca, N.Y., 1977, 287.
12. **Bryant, M. P.,** Microbial methane production — theoretic aspects, *J. Anim. Sci.,* 48, 193, 1979.
13. **Bryant, M. P. and Burkey, L. A.,** Cultural methods and some characteristics of some of the more numerous groups of bacteria in the bovine rumen, *J. Dairy Sci.,* 36, 205, 1953.
14. **Bryant, M. P., Tzeng, S. F., Robinson, I. M., and Joyner, A. E.,** Nutrient requirements of methanogenic bacteria, *Advances in Chemistry, Series 105,* Gould, A. G., Ed., American Chemical Society, Washington, D. C., 1971, 33.
15. **Buchanan, R. E. and Gibbons, N. E.,** *Bergey's Manual of Determinative Bacteriology,* 8th ed., Williams & Wilkins, Baltimore, Md., 1974.
16. **Caldwell, D. R. and Hudson, R. F.,** Sodium, an obligate requirement for predominant rumen bacteria, *Appl. Microbiol.,* 27, 549, 1974.
17. **Caldwell, D. R., White, D. C., Bryant, M. P., and Doetsch, R. N.,** Specificity of the heme requirement for the growth of *Bacteroides ruminicola, J. Bacteriol.,* 90(6), 1645, 1965.
18. **Chen, Y. R. and Hashimoto, A. G.,** Substrate utilization kinetic model for biological treatment processes, in *Biotechnology and Bioengineering,* Vol. 22, John Wiley & Sons, New York, 1980, 2081.
19. **Cheng, K. J. and Costerton, J. W.,** Adherent rumen bacteria — their role in the digestion of plant material, urea and epithelial cells, in *Digestive Physiology and Metabolism in Ruminants, Ruckebusch, Y. and Thivend, P.,* Eds., AVI, Westport, Conn., 1980, 227.
20. **Cheng, K. J., Stewart, C. S., Dinsdale, D., and Costerton, J. W.,** Electron microscopy of bacteria involved in the digestion of plant cell walls, *Ann. Feed Sci. Technol.,* 10, 93, 1984.

21. **Chynoweth, D. P. and Mah, R. A.,** Volatile acid formation in sludge digestion, *Adv. Chem. Ser.,* 105, 41, 1971.
22. **Converse, J. C., Graves, R. E., and Evans, G. W.,** Anaerobic degradation of dairy manure under mesophilic and thermophilic temperatures, *TASAE,* 20, 336, 1977.
23. **Cortois, D. E.,** Kinetics of bacterial growth: relationship between population density and specific growth rate of continuous cultures, *J. Gen. Microbiol.,* 21, 40, 1959.
24. **Cournoyer, M. S., Delisle, U., Ferland, D., and Chapnon, R.,** A mixed plug flow anaerobic digester for dairy manure, ASAE paper No. 84-4562, American Society of Agricultural Engineers, St. Joseph, Mich., 1984.
25. **Fanin, K. F. and Biljetina, R.,** Reactor designs for biomass digestion, in *Anaerobic Digestion of Biomass: Status Summary and R & D Needs,* Gas Research Institute, Chicago, 1983.
26. **Fischer, J. R. and Iannotti, E. L.,** Anaerobic digestion of manure from swine fed various diets, *Agric. Wastes,* 3, 201, 1981.
27. **Fischer, J. R., Iannotti, E. L., and Fulhage, C. D.,** The engineering, economics and management of a swine manure digester, Proceedings Methane Technology for Agriculture, Northeast Regional Agricultural Engineering Service Cornell University, Ithaca, N.Y., 1981.
28. **Fischer, J. R., Osburn, D. D., Meador, N. F., and Fulhage, C. D.,** Economics of a swine manure anaerobic digester, *Trans. ASAE,* 24(5), 1306, 1981.
29. **Fischer, J. R., Meador, N. W., Fulhage, C. D., and Harris, F. D.,** Energy self-sufficient swine production system-A model, *Trans. ASAE,* 24(5), 1264, and 1272, 1981.
30. **Fischer, J. R., Iannotti, E. L., and Porter, J. H.,** Anaerobic digestion of swine manure at various influent solids concentrations, *Agric. Wastes,* 11, 157, 1984.
31. **Fischer, J. R., Meador, N. F., Sievers, D. M., Fulhage, C. D., and Iannotti, E. L.,** Design and operation of a farm anaerobic digester for swine manure, *Trans. ASAE,* 22(5), 1129, 1979.
32. **Fischer, J. R., Iannotti, E. L., Porter, J. H., and Garcia, A.,** III, Producing methane gas from swine manure in a pilot-size digester, *Trans. ASAE,* 22(2), 370, 1979.
33. **Fischer, J. R., Iannotti, E. L., Stahl, T., Garcia, A., III, and Harris, F. D.,** Bioconversion of animal manure into electricity and a liquid fuel, *Biotechnol. Bioeng.,* 13, 527, 1983.
34. **Fischer, J. R., Gebhardt, M. R., and Fulhage, C. D.,** Effluent from a Swine Manure Anaerobic Digester as a Fertilizer for Corn, ASAE paper No. 84-2112, American Society of Agricultural Engineers, St. Joseph, Mich., 1984.
35. **Gaudy, A. F. and Gaudy, E. T.,** *Microbiology for Environmental Scientists and Engineers,* McGraw-Hill, New York, 1980.
36. **Georgacakis, D., Sievers, D. M., and Iannotti, E. L.,** Buffer stability in manure digesters, *Agric. Wastes,* 4, 427, 1982.
37. **Ghosh, S., Conrad, J. R., and Klass, D. L.,** Anaerobic acidogenesis of wastewater sludge, *J. Water Pollut. Control Fed.,* 47, 30, 1975.
38. **Gilman, F. E. and Bennett, S.,** Northern New England's Dairy Manure Digesters, ASAE Paper No. 84-4558, American Society of Agricultural Engineers, St. Joseph, Mich., 1984.
39. **Gosh, S. and Pohland, F. G.,** Kinetics of substrate assimilation and product formation in anaerobic digestion, *J. Water Pollut. Control Fed.,* 46, 749, 1974.
40. **Gomez-Alarcon, R., O'Dowd, C., Leedle, J. A. Z., and Bryant, M. P.,** 1, 4-Naphthoquinone and other nutrient requirements of *Succinivibrio dextrinosolvens, Appl. Environ. Microbiol.,* 44, 346, 1982.
41. **Goodrich, P. A., Kalis, S. M., Horvath, N. J., Nielsen, J. D., and Larson, V.,** Experiences in Building and Operating a Field Scale Anaerobic Digester, ASAE NCR-76-402, American Society of Agricultural Engineers, St. Joseph, Mich., 1976.
42. **Hasheider, R. J. and Sievers, D. M.** Limestone bed anaerobic filter for swine manure — laboratory study, *Trans. ASAE,* 27(3), 834, 1984.
43. **Hashimoto, A. G., Chen, Y. R., Varel, V. H., and Prior, R. L.,** Anaerobic fermentation of agricultural residues, in *Utilization and Recycle of Agricultural Residues,* Shuler, M. L., Ed., CRC Press, 1982.
44. **Hein, M. E., Smith, R. J., and Vetter, R. L.,** Some mechanical aspects of anaerobic digestion of beef manure, *ASAE,* 77, 4056, 1977.
45. **Herbeck, J. and Bryant, M.,** Nutritional features of the Intestinal Anaerobe *Ruminococcus bromii.* AM 28(6):1018. 1974.
46. **Hill, D. T.,** Design parameters and operating characteristics of animal waste anaerobic digestion systems — swine and poultry, *Agric. Wastes,* 5, 157, 1983.
47. **Hill, D. T.,** A comprehensive dynamic model for Animal Waste Methonogenesis, *Trans. ASAE,* 1274, 1982.
48. **Hobson, P. N., Bousfield, S., and Summers, R.,** *Methane Production from Agricultural and Domestic Wastes,* Halsted Press, New York, 1981.
49. **Hobson, P. N. and Shaw, B. G.,** The anaerobic digestion of waste from an intensive pig unit, *Water Res.,* 7, 437, 1973.

50. **Hobson, P. N. and Shaw, B. G.,** The bacterial population of piggery waste anaerobic digesters, *Water Res.,* 8, 507, 1974.

51. **Hobson, P. N. and Shaw, B. G.,** Inhibition of methane production by *Methanobacterium formicicum, Water Res.,* 10, 849, 1976.

52. **Holden, P. J., Speer, V. C., Steveers, E. J., and Zimmerman, D. R.,** Life cycle swine nutrition, Iowa State Univ. Coop. Ext. Serv. Pam 489 (Rev). 1976.

53. **Hungate, R. E.,** *The Rumen and its Microbes,* Academic Press, New York, 1966.

54. **Hungate, R. E.,** The rumen microbial ecosystem, *Annu. Rev. Ecol. Syst.,* 6, 39, 1975.

55. **Hungate, R. E. and Stack, R. J.,** Phenylpropionic acid; growth factor for *Ruminococcus albus, Appl. Environ. Microbiol.,* 44, 79, 1982.

56. **Iannotti, E. L.,** Microbiology and biochemistry of anaerobic digestion, Solar and Biomass Workshop, Atlanta, Georgia, 1982.

57. **Iannotti, E. L. and Fischer, J. R.,** Effects of ammonia, volatile acids, pH and sodium on growth of bacteria isolated from a swine manure digester, *Dev. Ind. Microbiol.,* 25, 741, 1984.

58. **Iannotti, E. L. and Fischer, J. R.,** Short term effect of antibiotics and feed additives on anaerobic digestion of swine manure, Annual meeting American Society of Microbiology, Dallas, Texas, 1981.

59. **Iannotti, E. L. and Fischer, J. R.,** Nutrition of bacteria from an anaerobic swine manure digester, Annual Meeting American Society of Microbiology, Atlanta, Georgia, 1982.

60. **Iannotti, E. L., Fischer, J. R., and Sievers, D. M.,** Media for the enumeration and isolation of bacteria from a swine waste digester, *Appl. Environ. Microbiol.,* 36(4), 555, 1978.

61. **Iannotti, E. L., Fischer, J. R., and Sievers, D. M.,** Characterization of bacteria from a swine manure digester, *Appl. Environ. Microbiol.,* 43, 136, 1982.

62. **Iannotti, E. L., Fischer, J. R., and Sievers, D. M.,** Medium for enhanced growth of bacteria from a swine manure digester, *Appl. Environ. Microbiol.,* 43, 247, 1982.

63. **Iannotti, E. L., Kafkewitz, D., Wolin, M. J., and Bryant, M. P.,** Glucose fermentation products of *Ruminococcus albus* grown in continuous culture with *Vibrio succinogenes:* changes caused by interspecies transfer of H_2, *J. Bacteriol.,* 113, 1231, 1973.

64. **Iannotti, E. L., Mueller, R. E., Fischer, J. R., and Sievers, D.,** Changes in a swine manure anaerobic digester with time after loading, Solar and Biomass Workshop, Atlanta, Georgia, 1983.

65. **Iannotti, E. L., Porter, J. H., Fischer, J. R., and Sievers, D. M.,** Changes in swine manure during anaerobic digestion, *Dev. Ind. Microbiol.,* 20, 519, 1979.

66. **Iannotti, E. L., Wulfers, M. K., Fischer, J. R., and Sievers, D. M.,** The effect of digester fluid, swine manure extract, rumen fluid and modified digester fluid on the growth of bacteria from an anaerobic digester, *Dev. Ind. Microbiol.,* 22, 565, 1981.

67. **Jarrell, K. F. and Sprott, G. D.,** Nickel transport in *Methanolbacterium bryantii, J. Bacteriol.,* 151(3), 1195, 1982.

68. **Jeris, J. S. and McCarty, P. L.,** The biochemistry of methane fermentation using ^{14}C tracers, *J. Water Pollut. Control. Fed.,* 37, 178, 1965.

69. **Jewell, W. J., Kabrick, R. M., Dell'Orto, S., Fanfoni, K. J., Cummings, R. J.,** Earthen-supported plug flow reactor for dairy applications, in *Proceedings from the Methane Technology for Agriculture Conference,* Cornell University, Ithaca, N.Y., 1981.

70. **Kaspar, H. F. and Wuhrmann,** Kinetic parameters and relative turnovers of some important catabolic reactins in digesting sludge, *Appl. Environ. Microbiol.,* 36, 1, 1978.

71. **Kirsch, E. J.,** Studies on the enumeration and isolation of obligate anaerobic bacteria from digesting sewage sludge, *Dev. Ind. Microbiol.,* 10, 170, 1969.

72. **Kirsch, E. J. and Sykes, R. M.,** Anaerobic digestion in biological waste treatment, *Progr. Ind. Microbiol.,* 9, 155, 1971.

73. **Klass, D. L.,** Methane from anaerobic fermentation, *Science,* 223, 1021, 1984.

74. **Kroeker, E. J., Lapp, H. M., Schulte, D. D., Haliburton, J. D., and Sparling, A. B.,** Methane production from animal wastes. II. Process stability, *Can. Soc. Agric. Eng.,* 76, 208, 1976.

75. **Kroeker, E. J., Schulte, D. D., Sparling, A. B., and Lapp, H. M.,** Anaerobic treatment process stability, *J. Water Pollut. Control Fed.,* 51, 718, 1979.

76. **Kugelman, I. J. and Chin, K. K.,** Toxicity, synergism and antagonism in anaerobic waste treatment processes, in *Anaerobic Biological Treatment Process,* Gould, R. F., Ed., Advance in Chemistry, Series 105, American Chemical Society, Washington, D. C., 1971, 55.

77. **Lapp, H. M., Schulte, D. D., Krocher, E. J., Sparling, A. B., and Topnik, B. H.,** Start-up of pilot scale swine manure digesters for methane production, in *Proceedings of 3rd. International Symposium on Livestock Waste,* American Society of Agriculturel Engineers, St. Joseph, Mich., 1975.

78. **Lev. M.,** The growth promoting activity of compounds of the vitamin K group and analogues for a rumen strain of *Fusiformis nigrescens, J. Gen. Microbiol.,* 20, 697, 1959.

79. **Linehan, B., Scheifinger, C. C., and Wolin, M. J.,** Nutritional requirements of *Selenomonas ruminantium* for growth on Lactate, Glycerol, or Glucose, *Appl. Environ. Microbiol.,* 35(2), 317, 1978.

80. **Lo, K. V. and Liao, P. H.,** Two-Phase Anaerobic Digestion of Screened Dairy Manure, ASAE paper No. 84-4560, American Society of Agricultural Engineers, St. Joseph, Mich., 1984.

81. **Mackie, R. I. and Bryant, M. P.,** Metabolic activity of fatty acid-oxizing bacteria and the contribution of acetate, propionate, butyrate and CO_2 to methanogenesis in cattle waste at 40 and 60°C, *Appl. Environ. Microbiol.,* 41, 1363, 1981.

82. **Mah, R. A., Smith, M. R., and Baresi, L.,** Studies on an acetate-fermenting strain of *Methanosarcina, Appl. Environ. Microbiol.,* 35, 1174, 1978.

83. **Mah, R. A. and Sussman, C.,** Microbiology of anaerobic sludge fermentation. I. Enumeration of the nonmethanogenic anaerobic bacteria, *Appl. Microbiol.,* 16, 358, 1968.

84. **Mah, R. A., Ward, D. M., Baresi, L., and Glass, T. L.,** Biogenesis of methane, *Ann. Rev. Microbiol.,* 31, 309, 1977.

85. **McCarty, P. L.,** Anaerobic waste treatment fundamentals. III. Toxic materials and their controls, *Public Works,* 95, 91, 1964.

86. **McCarty, P. L. and Brosseau, M. H.,** Effects of high concentration of individual VFA on anaerobic treatment, Proc. 18th Indust. Waste Conference, Purdue, 1963, 283.

87. **McCarty, P. L. and Mckinney, R. E.,** Volatile acid toxicity in anaerobic digestion, *J. Water Pollut. Control Fed.,* 33, 223, 1961.

88. **McCarty, P. L. and McKinney, R. E.,** Salt toxicity and anaerobic digestion, *J. Water Pollut. Control Fed.,* 33, 399, 1961.

89. **McInerney, M. J. and Bryant, M. P.,** Review of methane fermentation fundamentals, in *Fuel Gas Production from Biomass,* Volume II, Wise, D. L., Ed., CRC Press, Boca Raton, 1981, 19.

90. **McInerney, M. J., Bryant, M. P., Hespell, R. B., Costerton, J. W.,** *Syntrophomonas wolfei* gen nov. sp. nov., an anaerobic syntrophic fatty acid oxidizing bacterium, *Appl. Environ. Microbiol.,* 41, 1029, 1981.

91. **Mills, P. J.,** Minimization of energy input requirements of an anaerobic digester, *Agric. Wastes,* 1(1), 57, 1979.

92. **Monod, J.,** The growth of bacterial cultures, *Ann. Rev. Microbiol.,* 3, 371, 1949.

93. **Moore, W. E. C. and Holdeman, L. V.,** Human fecal flora; the normal flora of 20 Japanese Hawaiians, *Appl. Microbiol.,* 27, 961, 1974.

94. **Mountfort, D. O. and Asher, R. A.,** Changes in proportions of acetate and carbon dioxide used as methane precursors during the anaerobic digestion of bovine waste, *Appl. Environ. Microbiol.,* 36, 648, 1978.

95. **Mountfort, D. O. and Asher, R. A.** Effect of inorganic sulfide on the growth and metabolism of *Methanosarcina barkeri* strain DM, *Appl. Environ. Microbiol.,* 37(4), 670, 1979.

96. **Novak, J. T. and Carlson, D. A.,** The kinetics of anaerobic long-chain fatty acid degradation, *J. Water Pollut. Control Fed.,* 42, 1932, 1970.

97. **O'Roucke, J. R.,** Kinectics of Anaerobic Treatment at Reduced Temperatures, Ph.D. thesis, Stanford University, Standford, Calif., 1968.

98. **Patel, G. B., Breuil, C., and Agnew, B. J.,** Sulfur requirements for growth of *Acetivibrio cellulolyticus, Can. J. Microbiol.,* 28(7), 772,

99. **Person, S., Barlett, H. D., Regan, R. W., and Branding, A. E.,** Experiences from operating a full size anaerobic digester, in *Food Fertilizer and Agricultural Residues,* Lockr, R. C., Ed., Ann Arbor Science Publishers, Ann Arbor, Mich., 1977, 373.

100. **Pfeffer, J. T.,** Anaerobic fermentation, *Biotechnol. Bioeng.,* 16, 771, 1974.

101. **Pigg, D. L. and Veter, R. L.,** Dairy Cow Manure Digester and Cogenerator Performance, ASAE 84-4557, American Society of Agricultural Engineers, St. Joseph, Mich., 1984.

102. **Pigg, D. L.,** Commercial Size Anaerobic Digester Performance with Dairy Manure, ASAE 77-4055, American Society of Agricultural Engineers, St. Joseph, Mich., 1977.

103. **Pittman, K. A. and Bryant, M. P.,** Peptides and other nitrogen sources for growth of *Bacteroides ruminicola, J. Bacteriol.,* 88, 401, 1964.

104. **Poels, J., Van Assche, P., and Verstraete, W.,** Effects of disinfectants and antibiotics on the anaerobic digestion of piggery waste, *Agric. Wastes,* 9, 239, 1984.

105. **Rockey, D. A., Turnacliff, W., and Smith, R. J.,** A 1900 m³ Digester for Laying Hen Manure, Iowa, ASAE 78-4569, American Society of Agricultural Engineers, St. Joseph, Mich., 1978.

106. **Russell, J. B. and Baldwin, R. L.,** Comparison of substrate affinities among several rumen bacteria: a possible determinant of rumen bacterial competition, *Appl. Environ. Microbiol.,* 37, 531, 1979.

107. **Russell, J. B., Sharp, W. M., and Baldwin, R. L.,** The effect of pH on maximum bacterial growth rate: its possible role as a determinant of bacterial competition in the rumen, *J. Anim. Sci.,* 48, 251, 1979.

108. **Saddler, J. N. and Khan, A. W.,** Cellulose degradation by a new isolate from sewage sludge, a member of *Bacteriodaceae* family, *Can. J. Microbiol.,* 25, 1427, 1979.

109. **Salanitro, J. P., Blake, I. G., and Muirhead, P. A.,** Isolation and identification of fecal bacteria from adult swine, *Appl. Environ. Microbiol.,* 33, 79, 1977.

110. **Sievers, D. M. and Brune, D. E.,** Carbon/nitrogen ratio and anaerobic digestion of swine waste, *Trans. ASAE*, 21, 537, 1978.

110b. **Sievers, D. M., Huff, H. E., and Iannotti, E. L.,** Potential methane production of the clarified liquids from a settling basin, *Livestock Waste: A Renewable Resource*, 1980, 98.

111. **Smith, C. J. and Iannotti, E. L.,** Enumeration and characterization of heterotrophic bacteria from anaerobic marine sediments, *Dev. Ind. Microbiol.*, 25, 727, 1984.

112. **Smith, P. H. and Mah, R.,** Kinetics of acetate metabolism during sludge digestion, *Appl. Microbiol.*, 14(3), 368, 1966.

113. **Smith, R. J.,** personal communications, Department of Agricultural Engineering, Iowa State University, Ames, Iowa, 1978.

114. **Spoelstra, S. F.,** Enumeration and isolation of anaerobic microbiota of piggery wastes, *Appl. Environ. Microbiol.*, 35, 841, 1978.

115. **Stahl, T., Harris, F. D., and Fischer, J. R.,** Farm Scale Biogas Fueled Engine Induction Generator System, ASAE 82-3543, American Society of Agricultural Engineers, St. Joseph, Mich., 1982.

116. **Sykes, R. M. and Kirsch, E. J.,** Accumulation of methanogenic substances in CCl_4 inhibited anaerobic sewage sludge digester cultures, *Water Res.*, 6, 41, 1972.

117. **te Boekhost, R. H., Ogilvie, J. R., and Pos, J.,** An Overview of Current Simulation Models for an Anaerobic Digester, Proc. 4th. Int. Symp. Livestock Wastes, American Society of Agricultural Engineers, St. Joseph, Mich., 1980.

118. **Thiel, P. G.,** The effect of methane analogues on methanogenesis in anaerobic digestion, *Water Res.*, 38, 215, 1969.

119. **Toerien, D. F.,** Population description of the non-methanogenic phase of anaerobic digestion. I. Isolation, characterization and identification of numerically important bacteria, *Water Res.*, 4, 129, 1970.

120. **Toerien, D. F. and Hattingh, W. H. J.,** Anaerobic digestion. I. The microbiology of anaerobic digestion, *Water Res. (G.B.)*, 3, 1969.

121. **Turnacliff, W. and Custer, L.,** Evaluation of Methane Gas System and Design for a Hog Farm, Final report to the state of Colorado, Office of Energy Conservation, 1978.

122. **Van der Berg, L., Patel, G. B., Clark, D. S., and Lentz, C. P.,** Factors affecting rate of methane formation from acetate by enrichment cultures, *Can. J. Microbiol.*, 22, 1312, 1976.

123. **Van der Berg, L. et al.,** Effects of sulphate, iron and hydrogen on the microbiological conversion of acetic acid to methane, *J. Appl. Bacteriol.*, 48, 437, 1980.

124. **Van Soest, P. J.,** Physico-chemical aspects of fibre digestion, in *Digestion and Metabolism in the Ruminant*, McDonald, I. W. and Warner, A. C. I., Eds., University of New England Publishing Unit, Armidale, Australia, 1975, 351.

125. **Van Velsen, A. F. M.,** Anaerobic digestion of piggery waste. I. The influence of detention time and manure concentration, *Neth. J. Agric. Sci.*, 25, 151, 1977.

126. **Van Velsen, A. F. M.,** Anaerobic digestion of piggery waste. III. Influence of temperature, *Neth. J. Agric. Sci.*, 27(4), 255, 1977.

127. **Van Velsen, A. F. M.,** Adaptation of methanogenic sludge to high ammonia-nitrogen concentration, *Water Res.*, 13(109), 995, 1979.

128. **Varel, V. H.,** Methane-production fermenter, *Microb. Ecol.*, 10, 15, 1984.

129. **Varel, V. H. and Bryant, M. P.,** Nutritional features of *Bacteroides fragilis* subspecies *fragilis*, *Appl. Microbiol.*, 28(2), 251, 1974.

130. **Varel, V. H. and Hashimoto, A. G.,** Effect of dietary monensin or chlortetracycline on methane production from cattle waste, *Appl. Environ. Microbiol.*, 41, 29, 1981.

131. **Varel, V. H. and Hashimoto, A. G.,** Methane production by fermentor cultures acclimated to waste from cattle fed monensin, lasalocid, salinomycin, or avoparcin, *Appl. Environ. Microbiol.*, 44(6), 1415, 1982.

132. **Verhoff, F. H. and Sundereson, K. R.,** A mechanism of microbial cell growth, *Biotechnol. Bioeng.*, 14, 411, 1972.

133. **Weeks, S. A.,** Complete systems — challenges and responsibilities of the supplier, Proceedings Methane Technology for Agriculture Cornell University, Ithaca, N. Y., 1981.

134. **Wellinger, A. and Kaufmann, R.,** Psychrophyllic methane generation form pig manure, *Process Biochem.*, Sept/Oct., 1982.

135. **Welsh, F. W., Schulte, D. D., Krocker, E. J., and Lapp, H. M.,** The effect of anaerobic digestion on swine manure odor, *Can. Agric. Eng.*, 19(2), 1977.

136. **Wofe, R. S.,** Microbiol biochemistry of methane — a study in contrast. I. Methanogenesis. II. Methanotrophy, *Microbial Biochemistry*, Vol. 21, Quayle, J. R., Ed., 1979.

137. **Wohlt, J. E., Frobish, R. A., Charron, R. E., Davis, C. L., and Bryant, M. P.,** Chemical components and methane production from dairy wastes, *Am. Soc. Anim. Sci. Abstr.*, 1978.

138. **Wolin, M. J.,** Interactions between the bacterial species of the rumen, in *Digestion and Metabolism in the Ruminant*, McDonald, I. W. and Warner, A. C. I., Eds., University of New England Publishing Unit, Armidale, Australia, 1975, 134.

139. **Wolin, M. J.,** Hydrogen transfer in Microbial communities, in *Microbial Interactions and Communities,* Bull, A. T. and Slater, J. H., Eds., Vol. 1, Academic Press, London, 1982, 323.
140. **Wozny, M. A., Bryant, M. P., Holderman, L. V., and Moore, W. E. C.,** Urease assay and urease-producing species of anaerobes in the bovine rumen and human feces, *Appl. Environ. Microbiol.,* 33, 1097, 1977.
141. **Zeikus, J. G.,** The biology of methanogenic bacteria, *Bacteriol. Rev.,* 41(2), 514, 1977.
142. **Ianotti, E. L. and Fischer, J. R.,** unpublished data.

APPENDIX A

UNITS AND CONVERSION FACTORS

The fundamental system of units used in scientific work world-wide is "System International d'Unites" or SI (metric). However, in the U.S., both the SI and the U.S. Customary Units (USCS) are used. Therefore, the conversion factors for frequently used units and the prefixes used in the SI system are given below.

Commonly Used SI Units Prefixes

Prefix	Symbol	Multiplying factor
Tera	T	10^{12}
Giga	G	10^{9}
Mega	M	10^{6}
Kilo	k	10^{3}
Hecto	h	10^{2}
Deci	d	10^{-1}
Centi	c	10^{-2}
Milli	m	10^{-3}
Micro	μ	10^{-6}
Nano	n	10^{-9}

Some Commonly Used Unit Conversions

Length

1 in.	= 2.54 cm		1 cm	= 0.3937 in.
1 ft	= 0.3048 m		1 m	= 3.2808 ft
1 yd	= 0.9144 m		1 km	= 0.6215 mi
1 mi	= 1.609 km			

Area

1 in.^{2}	= 6.452 cm^2		1 cm^2	= 0.155 in.^{2}
1 ft^2	= 0.093 m^2		1 m^2	= 10.752 ft^2
1 acre	= 4840 yd^2		1 ha	= 2.4691 acres
1 acre	= 4047 m^2			

Volume

1 ft^3	= 0.0283 m^3
	= 28.32 ℓ
1 ft^3	= 7.48 gal (U.S.)
1 gal (U.S.)	= 3.785 ℓ
1 gal (U.K.).	= 4.546 ℓ
1 bbl	= 42 gal (U.S.)

Volumetric Rate

1 cfm (ft^3/min)	= 0.472 ℓ/sec
1 gal/min (GPM)	= 0.0631 ℓ/sec

Mass

1 lb	= 0.4536 kg
1 ton	= 2240 lb
1 ton (metric)	= 1000 kg

Mass Flow Rate

1 lb/hr	= 0.000126 kg/sec

UNITS AND CONVERSION FACTORS (continued)

Specific Volume

$$1 \text{ ft}^3/\text{lb} = 0.0624 \text{ m}^3/\text{kg}$$

Density

$$1 \text{ lb/ft}^3 = 16.0185 \text{ kg/m}^3$$

Force, Pressure

1 lbf	= 4.448 N	1 N/m^2	= 1 Pa
1 lbf/in.^{2} (psi)	= 6894.8 N/m^2		= 0.000145 psi
1 in. Hg	= 3.386 kPa	1 atm (standard)	= 14.676 psi
			= 101.325 kPa

Energy

1 Btu	= 1.055 J	1 Btu/lb	= 2.326 kJ/kg
1 Therm	= 105.506 MJ	1 Btu/ft^3	= 37.259 kJ/m^3
1 kWh	= 3.6 MJ	1 Btu/ft^2	= 11.36 kJ/m^2
1 Cal	= 4.1868 J	1 Btu/lb °F	= 4.187 kJ/kg °C
1 Langley	= 41.86 kJ/m^2	1 Cal/cm^2	= 41.87 MJ/m^2

Power, Energy Flux

1 Btu/hr	= 0.2931 W	1 Btu/hr ft^2	= 3.1547 W/m^2
1 kcal/hr	= 1.163 W	1 Btu/hr ft^2 °F	= 5.678 W/m^2 °C
1 hp	= 0.7457 kW	1 Btu/hr ft °F	= 1.7031 W/m^2 °C
1 W/ft^2	= 10.76 W/m^2	1 Langley/hr	= 697.4 W/m^2
1 ton (refrigeration)	= 3.517 kW		

Temperature

Scales		**Differences**
°F	= °C × 1.8 + 32	1 °F = 0.55556 °C
°C	= (F − 32) × 5/9	1 °C = 1.8 °F
K	= °C + 273	
R	= °F + 462	

APPENDIX B

TABLES OF SOLAR RADIATION DATA IN U.S. CONVENTIONAL UNITS AND TABLES OF SUN ANGLES

Table 1
AVERAGE DAILY TOTAL SOLAR RADIATION ON SOUTH-FACING SURFACES IN NORTHERN HEMISPHERE

Latitude = 0°N

Month	Horiz	15	30	45	60	75	90
1	2741	3008	3095	2998	2723	2288	1723
2	2850	2987	2947	2734	2361	1855	1249
3	2886	2838	2625	2262	1773	1192	597
4	2793	2567	2197	1709	1143	580	438
5	2641	2298	1834	1290	735	434	453
6	2540	2153	1657	1105	587	447	459
7	2574	2210	1732	1188	657	456	468
8	2696	2421	2015	1509	953	492	468
9	2816	2689	2409	1996	1479	897	470
10	2836	2892	2779	2506	2091	1562	955
11	2761	2979	3021	2883	2577	2122	1549
12	2689	2987	3108	3043	2796	2384	1835

Note: Clearness no. = 1.0; ground reflection = 0.2. All units in British thermal units per square foot.

Table 2
AVERAGE DAILY TOTAL SOLAR RADIATION ON SOUTH-FACING SURFACES IN NORTHERN HEMISPHERE

Latitude = 10°N

Month	Horiz	15	30	45	60	75	90
1	2396	2755	2949	2966	2804	2474	1999
2	2611	2858	2935	2837	2570	2154	1617
3	2805	2875	2777	2517	2113	1593	993
4	2882	2764	2488	2074	1555	977	496
5	2861	2606	2209	1705	1141	612	485
6	2821	2508	2063	1530	967	509	501
7	2831	2545	2124	1606	1046	558	506
8	2847	2668	2341	1892	1359	802	491
9	2813	2796	2620	2295	1846	1303	719
10	2665	2832	2833	2669	2350	1898	1344
11	2454	2769	2920	2896	2700	2344	1852
12	2317	2701	2924	2970	2837	2533	2079

Note: Clearness no. = 1.0; ground reflection = 0.2. All units in British thermal units per square foot.

Table 3
AVERAGE DAILY TOTAL SOLAR RADIATION ON SOUTH-FACING SURFACES IN NORTHERN HEMISPHERE

Latitude = 20°N

Month	Horiz	15	30	45	60	75	90
1	1980	2408	2691	2811	2760	2540	2166
2	2287	2628	2812	2828	2673	2359	1906
3	2629	2812	2830	2681	2376	1936	1390
4	2877	2872	2702	2380	1930	1387	807
5	2996	2842	2531	2091	1558	990	537
6	3023	2801	2429	1943	1386	832	533
7	3005	2813	2470	2005	1461	904	536
8	2907	2833	2601	2229	1747	1196	662
9	2714	2808	2741	2518	2153	1672	1111
10	2403	2671	2782	2728	2514	2154	1672
11	2071	2462	2708	2790	2703	2454	2058
12	1880	2324	2628	2773	2747	2554	2205

Note: Clearness no. = 1.0; ground reflection = 0.2. All units in British thermal units per square foot.

Table 4
AVERAGE DAILY TOTAL SOLAR RADIATION ON SOUTH-FACING SURFACES IN NORTHERN HEMISPHERE

Latitude = 30°N

Month	Horiz	15	30	45	60	75	90
1	1515	1977	2321	2522	2568	2454	2190
2	1891	2303	2577	2695	2648	2441	2086
3	2362	2647	2776	2740	2541	2194	1722
4	2774	2882	2824	2605	2241	1759	1198
5	3038	2992	2781	2422	1946	1394	836
6	3137	3017	2734	2316	1797	1231	707
7	3090	3001	2750	2358	1860	1302	765
8	2873	2907	2779	2498	2086	1577	1019
9	2520	2720	2762	2645	2377	1975	1469
10	2062	2412	2620	2672	2564	2304	1909
11	1630	2069	2384	2554	2567	2424	2132
12	1401	1867	2221	2439	2506	2416	2177

Note: Clearness no. = 1.0; ground reflection = 0.2. All units in British thermal units per square foot.

Table 5
AVERAGE DAILY TOTAL SOLAR RADIATION ON SOUTH-FACING SURFACES IN NORTHERN HEMISPHERE

Latitude = 40°N

Month	Horiz	15	30	45	60	75	90
1	1024	1474	1834	2082	2199	2178	2021
2	1441	1890	2225	2424	2473	2370	2121
3	2014	2382	2609	2679	2589	2344	1960
4	2578	2792	2845	2735	2468	2066	1557
5	2988	3051	2945	2681	2280	1775	1213
6	3164	3148	2964	2629	2172	1632	1066
7	3085	3101	2950	2646	2214	1691	1129
8	2744	2884	2862	2681	2355	1910	1381
9	2239	2530	2675	2664	2498	2188	1756
10	1654	2059	2343	2485	2478	2320	2023
11	1154	1596	1943	2171	2264	2217	2032
12	906	1343	1700	1950	2079	2077	1943

Note: Clearness no. = 1.0; ground reflection = 0.2. All units in British thermal units per square foot.

Table 6
AVERAGE DAILY TOTAL SOLAR RADIATION ON SOUTH-FACING SURFACES IN NORTHERN HEMISPHERE

Latitude = 50°N

Month	Horiz	15	30	45	60	75	90
1	543	909	1220	1455	1598	1639	1575
2	960	1395	1747	1991	2110	2098	1953
3	1598	2020	2322	2485	2496	2355	2072
4	2295	2601	2757	2754	2591	2282	1847
5	2851	3015	3014	2852	2540	2104	1578
6	3110	3192	3108	2868	2489	2001	1447
7	2996	3112	3062	2854	2504	2039	1499
8	2528	2762	2842	2764	2534	2169	1698
9	1879	2240	2472	2559	2495	2284	1941
10	1197	1618	1942	2150	2225	2164	1970
11	671	1055	1376	1612	1747	1771	1684
12	436	767	1051	1270	1408	1456	1411

Note: Clearness no. = 1.0; ground reflection = 0.2. All units in British thermal units per square foot.

Table 7
AVERAGE DAILY TOTAL SOLAR RADIATION ON SOUTH-FACING SURFACES IN NORTHERN HEMISPHERE

Latitude = 60°N

Month	Horiz	15	30	45	60	75	90
1	141	312	463	586	671	712	708
2	483	826	1120	1344	1482	1527	1474
3	1130	1563	1904	2129	2223	2179	2001
4	1935	2310	2553	2648	2589	2379	2034
5	2643	2889	2984	2923	2708	2355	1890
6	2995	3157	3166	3022	2731	2314	1803
7	2843	3038	3081	2971	2712	2322	1833
8	2236	2544	2714	2733	2602	2329	1933
9	1453	1852	2145	2310	2339	2227	1983
10	718	1092	1402	1626	1749	1763	1667
11	238	465	663	821	926	972	956
12	73	181	279	359	415	445	446

Note: Clearness no. = 1.0; ground reflection = 0.2. All units in British thermal units per square foot.

Table 8
HOURLY SUN ANGLES FOR LATITUDE = 0°N

Solar time		Jan. 21		Feb. 21		March 21	
AM	PM	Altitude	Azimuth	Altitude	Azimuth	Altitude	Azimuth
12	0	69.9	0.0	79.1	0.0	89.9	0.0
11	1	65.1	35.3	71.5	53.4	75.0	89.5
10	2	54.4	53.8	58.3	69.0	60.0	89.7
9	3	41.6	62.6	44.0	74.8	45.0	89.8
8	4	28.0	67.1	29.4	77.5	30.0	89.8
7	5	14.1	69.2	14.7	78.7	15.0	89.9
6	6	0.0	69.9	0.0	79.1	0.0	89.9

		April 21		May 21		June 21	
		Altitude	Azimuth	Altitude	Azimuth	Altitude	Azimuth
12	0	78.5	180.0	70.0	180.0	66.6	180.0
11	1	71.2	128.3	65.2	144.6	62.4	149.2
10	2	58.1	112.2	54.5	126.0	52.6	130.9
9	3	43.9	106.1	41.6	117.2	40.4	121.5
8	4	29.3	103.3	28.0	112.8	27.3	116.6
7	5	14.7	101.9	14.1	110.6	13.7	114.2
6	6	0.0	101.5	0.0	110.0	0.0	113.4

		July 21		Aug. 21		Sept. 21	
		Altitude	Azimuth	Altitude	Azimuth	Altitude	Azimuth
12	0	69.3	180.0	77.6	180.0	88.9	180.0
11	1	64.7	145.5	70.6	130.4	75.0	94.1
10	2	54.1	127.0	57.7	113.8	60.0	92.1
9	3	41.4	118.1	43.7	107.3	45.0	91.5
8	4	27.9	113.5	29.2	104.3	30.0	91.2
7	5	14.0	111.3	14.6	102.9	15.0	91.1
6	6	0.0	110.7	0.0	102.4	0.0	91.1

		Oct. 21		Nov. 21		Dec. 21	
		Altitude	Azimuth	Altitude	Azimuth	Altitude	Azimuth
12	0	79.6	0.0	70.3	0.0	66.6	0.0
11	1	71.8	54.8	65.4	35.8	62.4	30.8
10	2	58.4	69.9	54.6	54.4	52.6	49.1
9	3	44.1	75.5	41.7	63.1	40.5	58.5
8	4	29.5	78.1	28.1	67.5	27.3	63.4
7	5	14.7	79.3	14.1	69.7	13.7	65.8
6	6	0.0	79.6	0.0	70.3	0.0	66.6

Note: To convert from solar time to local time, apply corrections for longitude and equation of time.

Table 9
HOURLY SUN ANGLES FOR LATITUDE = 10°N

| Solar time | | Jan. 21 | | Feb. 21 | | March 21 | |
AM	PM	Altitude	Azimuth	Altitude	Azimuth	Altitude	Azimuth
12	0	59.9	0.0	69.1	0.0	79.9	0.0
11	1	56.5	26.1	64.3	35.9	72.0	56.7
10	2	47.8	44.4	53.6	55.8	58.5	73.0
9	3	36.5	55.7	40.6	66.2	44.1	80.0
8	4	23.7	62.7	26.8	72.3	29.5	84.1
7	5	10.4	67.2	12.6	76.4	14.7	87.2
6	6	− 3.4	70.2	− 1.9	79.3	0.0	89.9

| | | April 21 | | May 21 | | June 21 | |
		Altitude	Azimuth	Altitude	Azimuth	Altitude	Azimuth
12	0	88.5	180.0	80.0	180.0	76.6	180.0
11	1	75.2	97.3	72.4	126.4	70.4	135.0
10	2	60.5	95.8	59.4	112.6	58.4	118.9
9	3	45.8	96.3	45.5	108.4	45.1	113.3
8	4	31.1	97.5	31.5	107.4	31.4	111.4
7	5	16.5	99.2	17.4	108.0	17.6	111.6
6	6	2.0	101.4	3.4	109.7	4.0	113.1

| | | July 21 | | Aug. 21 | | Sept. 21 | |
		Altitude	Azimuth	Altitude	Azimuth	Altitude	Azimuth
12	0	79.3	180.0	87.6	180.0	81.1	0.0
11	1	72.1	128.2	75.1	100.8	72.6	60.0
10	2	59.2	113.8	60.5	97.6	58.9	75.2
9	3	45.5	109.4	45.8	97.6	44.4	81.6
8	4	31.5	108.2	31.2	98.6	29.7	85.5
7	5	17.4	108.7	16.6	100.1	15.0	88.4
6	6	3.5	110.4	2.1	102.2	0.2	91.1

| | | Oct. 21 | | Nov. 21 | | Dec. 21 | |
		Altitude	Azimuth	Altitude	Azimuth	Altitude	Azimuth
12	0	69.6	0.0	60.3	0.0	56.6	0.0
11	1	64.8	36.7	56.8	26.4	53.5	23.5
10	2	53.9	56.5	48.1	44.8	45.5	40.9
9	3	40.8	66.8	36.7	56.1	34.7	52.1
8	4	26.9	72.9	23.9	63.1	22.5	59.3
7	5	12.7	76.9	10.5	67.6	9.5	64.0
6	6	− 1.8	79.8	− 3.4	70.6	− 4.0	66.9

Note: To convert from solar time to local time, apply corrections for longitude and equation of time.

Table 10
HOURLY SUN ANGLES FOR LATITUDE = 20°N

Solar time		Jan. 21		Feb. 21		March 21	
AM	PM	Altitude	Azimuth	Altitude	Azimuth	Altitude	Azimuth
12	0	49.9	0.0	59.1	0.0	69.9	0.0
11	1	47.3	21.0	55.8	26.9	65.1	37.9
10	2	40.3	38.0	47.3	46.3	54.4	59.2
9	3	30.4	50.4	36.0	59.1	41.6	71.0
8	4	18.9	59.3	23.4	67.9	28.0	78.7
7	5	6.4	65.9	10.0	74.4	14.0	84.6
6	6	−6.8	71.0	−3.7	79.8	0.0	89.9

		April 21		May 21		June 21	
		Altitude	Azimuth	Altitude	Azimuth	Altitude	Azimuth
12	0	81.5	0.0	90.0	0.0	86.6	180.0
11	1	73.3	61.8	75.9	92.5	75.7	106.6
10	2	60.0	78.2	61.8	95.2	62.0	102.5
9	3	46.0	86.0	47.8	98.1	48.2	103.2
8	4	31.9	91.4	33.9	101.2	34.5	105.3
7	5	17.9	96.1	20.2	104.7	21.1	108.3
6	6	3.9	100.9	6.7	108.9	7.8	112.2

		July 21		Aug. 21		Sept. 21	
		Altitude	Azimuth	Altitude	Azimuth	Altitude	Azimuth
12	0	89.3	0.0	82.4	0.0	71.1	0.0
11	1	75.9	95.3	73.7	64.5	66.1	39.6
10	2	61.9	96.6	60.3	79.9	55.1	60.9
9	3	47.9	99.0	46.3	87.2	42.1	72.4
8	4	34.1	101.9	32.2	92.4	28.4	80.0
7	5	20.4	105.4	18.1	97.0	14.5	85.8
6	6	6.9	109.5	4.2	101.7	0.4	91.0

		Oct. 21		Nov. 21		Dec. 21	
		Altitude	Azimuth	Altitude	Azimuth	Altitude	Azimuth
12	0	59.6	0.0	50.3	0.0	46.6	0.0
11	1	56.2	27.3	47.7	21.2	44.2	19.3
10	2	47.6	46.9	40.6	38.3	37.6	35.4
9	3	36.3	59.7	30.7	50.7	28.3	47.4
8	4	23.6	68.4	19.1	59.6	17.2	56.3
7	5	10.2	74.9	6.5	66.2	5.0	62.8
6	6	−3.5	80.3	−6.6	71.4	−7.8	67.8

Note: To convert from solar time to local time, apply corrections for longitude and equation of time.

Table 11
HOURLY SUN ANGLES FOR LATITUDE = 30°N

Solar time		Jan. 21		Feb. 21		March 21	
AM	**PM**	**Altitude**	**Azimuth**	**Altitude**	**Azimuth**	**Altitude**	**Azimuth**
12	0	39.9	0.0	49.1	0.0	59.9	0.0
11	1	37.9	17.9	46.6	21.7	56.7	28.1
10	2	32.2	33.7	39.9	39.8	48.5	49.0
9	3	23.8	46.5	30.5	53.7	37.7	63.3
8	4	13.6	56.8	19.3	64.3	25.6	73.8
7	5	2.2	65.2	7.2	73.0	12.9	82.3
6	6	−9.9	72.4	−5.4	80.5	−0.1	89.9

		April 21		May 21		June 21	
		Altitude	**Azimuth**	**Altitude**	**Azimuth**	**Altitude**	**Azimuth**
12	0	71.5	0.0	80.0	0.0	83.4	0.0
11	1	66.9	40.2	73.1	57.0	75.1	67.4
10	2	56.6	62.9	61.1	76.7	62.5	83.4
9	3	44.4	76.0	48.3	86.9	49.5	91.8
8	4	31.6	85.2	35.3	94.3	36.6	98.2
7	5	18.6	92.8	22.4	100.9	23.9	104.3
6	6	5.7	100.0	9.8	107.5	11.5	110.6
5	7	−6.9	107.6	−2.3	114.7	−0.4	117.6

		July 21		Aug. 21		Sept. 21	
		Altitude	**Azimuth**	**Altitude**	**Azimuth**	**Altitude**	**Azimuth**
12	0	80.7	0.0	72.4	0.0	61.1	0.0
11	1	73.6	58.8	67.6	41.5	57.7	29.0
10	2	61.4	78.0	57.1	64.2	49.4	50.2
9	3	48.5	87.8	44.9	77.0	38.4	64.5
8	4	35.6	95.0	32.0	86.1	26.3	74.9
7	5	22.7	101.5	19.1	93.6	13.5	83.3
6	6	10.2	108.1	6.2	100.8	0.5	90.9
5	7	−1.9	115.3	−6.4	108.3	****	98.6

		Oct. 21		Nov. 21		Dec. 21	
		Altitude	**Azimuth**	**Altitude**	**Azimuth**	**Altitude**	**Azimuth**
12	0	49.6	0.0	40.3	0.0	36.6	0.0
11	1	47.1	22.0	38.2	18.1	34.7	16.8
10	2	40.4	40.2	32.5	33.9	29.3	31.7
9	3	30.8	54.1	24.1	46.8	21.3	44.1
8	4	19.6	64.8	13.8	57.1	11.4	54.2
7	5	7.5	73.4	2.4	65.5	0.4	62.4
6	6	−5.2	81.0	−9.7	72.8	****	69.4

Note: To convert from solar time to local time, apply corrections for longitude and equation of time.

Table 12
HOURLY SUN ANGLES FOR LATITUDE = 40°N

Solar time		Jan. 21		Feb. 21		March 21	
AM	PM	Altitude	Azimuth	Altitude	Azimuth	Altitude	Azimuth
12	0	29.9	0.0	39.1	0.0	49.9	0.0
11	1	28.3	16.0	37.2	18.6	47.6	22.6
10	2	23.7	30.9	32.0	35.4	41.4	41.8
9	3	16.7	43.9	24.2	49.6	32.7	57.2
8	4	8.0	55.2	14.8	61.6	22.4	69.5
7	5	−2.0	65.2	4.2	72.0	11.3	80.1
6	6	****	74.3	−7.0	81.6	−0.1	89.9

		April 21		May 21		June 21	
		Altitude	Azimuth	Altitude	Azimuth	Altitude	Azimuth
12	0	61.5	0.0	70.0	0.0	73.4	0.0
11	1	58.6	29.1	66.2	37.1	69.2	41.9
10	2	51.1	51.3	57.5	60.9	59.8	65.8
9	3	41.2	67.1	46.8	76.0	48.8	80.2
8	4	30.3	79.2	35.4	87.2	37.4	90.7
7	5	18.8	89.4	24.0	96.6	25.9	99.7
6	6	7.4	98.9	12.7	105.6	14.8	108.4
5	7	−3.8	108.5	1.9	114.7	4.2	117.3

		July 21		Aug. 21		Sept. 21	
		Altitude	Azimuth	Altitude	Azimuth	Altitude	Azimuth
12	0	70.7	0.0	62.4	0.0	51.1	0.0
11	1	66.8	37.9	59.4	29.8	48.8	23.1
10	2	57.9	61.8	51.8	52.2	42.5	42.7
9	3	47.2	76.8	41.9	68.0	33.6	58.1
8	4	35.8	87.8	30.8	80.0	23.3	70.5
7	5	24.3	97.2	19.4	90.1	12.1	81.1
6	6	13.1	106.1	8.0	99.6	0.7	90.8
5	7	2.4	115.2	−3.2	109.1	****	100.6

		Oct. 21		Nov. 21		Dec. 21	
		Altitude	Azimuth	Altitude	Azimuth	Altitude	Azimuth
12	0	39.6	0.0	30.3	0.0	26.6	0.0
11	1	37.8	18.8	28.7	16.1	25.0	15.2
10	2	32.5	35.7	24.1	31.0	20.7	29.4
9	3	24.7	49.9	17.0	44.1	14.0	42.0
8	4	15.1	61.9	8.3	55.5	5.5	53.0
7	5	4.6	72.4	−1.7	65.5	−4.2	62.7

Note: To convert from solar time to local time, apply corrections for longitude and equation of time.

Table 13
HOURLY SUN ANGLES FOR LATITUDE = 50°N

| Solar time | | Jan. 21 | | Feb. 21 | | March 21 | |
AM	PM	Altitude	Azimuth	Altitude	Azimuth	Altitude	Azimuth
12	0	19.9	0.0	29.1	0.0	39.9	0.0
11	1	18.6	14.9	27.7	16.7	38.3	19.2
10	2	15.0	29.1	23.7	32.4	33.7	36.9
9	3	9.4	42.3	17.6	46.7	26.9	52.5
8	4	2.2	54.5	9.8	59.7	18.6	66.1
7	5	−6.1	65.8	1.1	71.6	9.5	78.3
6	6	****	76.8	−8.3	82.9	−0.1	89.9

| Solar time | | April 21 | | May 21 | | June 21 | |
AM	PM	Altitude	Azimuth	Altitude	Azimuth	Altitude	Azimuth
12	0	51.5	0.0	60.0	0.0	63.4	0.0
11	1	49.6	23.0	57.7	27.1	61.0	29.3
10	2	44.3	43.2	51.7	49.3	54.6	52.4
9	3	36.8	59.9	43.6	66.5	46.2	69.6
8	4	27.9	73.8	34.3	80.2	36.8	83.1
7	5	18.4	86.0	24.7	92.1	27.2	94.8
6	6	8.8	97.5	15.2	103.2	17.7	105.6
5	7	−0.6	108.8	6.1	114.1	8.7	116.3

| Solar time | | July 21 | | Aug. 21 | | Sept. 21 | |
AM	PM	Altitude	Azimuth	Altitude	Azimuth	Altitude	Azimuth
12	0	60.7	0.0	52.4	0.0	41.1	0.0
11	1	58.3	27.5	50.5	23.4	39.4	19.6
10	2	52.3	49.9	45.1	43.8	34.8	37.5
9	3	44.1	67.1	37.5	60.5	28.0	53.2
8	4	34.8	80.8	28.6	74.4	19.6	66.8
7	5	25.2	92.6	19.1	86.6	10.4	79.1
6	6	15.7	103.6	9.5	98.1	0.8	90.7
5	7	6.6	114.5	0.1	109.4	−8.7	102.3

| Solar time | | Oct. 21 | | Nov. 21 | | Dec. 21 | |
AM	PM	Altitude	Azimuth	Altitude	Azimuth	Altitude	Azimuth
12	0	29.6	0.0	20.3	0.0	16.6	0.0
11	1	28.2	16.8	19.0	14.9	15.4	14.3
10	2	24.2	32.6	15.4	29.2	11.9	28.0
9	3	18.0	47.0	9.8	42.5	6.5	40.8
8	4	10.3	60.0	2.5	54.7	−0.6	52.6
7	5	1.5	71.9	−5.8	66.1	−8.7	63.7

Note: To convert from solar time to local time, apply corrections for longitude and equation of time.

Table 14
HOURLY SUN ANGLES FOR LATITUDE = 60°N

Solar time		Jan. 21		Feb. 21		March 21	
AM	PM	Altitude	Azimuth	Altitude	Azimuth	Altitude	Azimuth
12	0	9.9	0.0	19.1	0.0	29.9	0.0
11	1	9.0	14.2	18.1	15.5	28.7	17.2
10	2	6.3	28.2	15.2	30.6	25.5	33.6
9	3	2.0	41.6	10.6	44.9	20.6	49.1
8	4	− 3.6	54.6	4.7	58.6	14.4	63.4
7	5	****	67.1	− 2.1	71.7	7.3	76.9

		April 21		May 21		June 21	
		Altitude	Azimuth	Altitude	Azimuth	Altitude	Azimuth
12	0	41.5	0.0	50.0	0.0	53.4	0.0
11	1	40.3	19.4	48.6	21.6	52.0	22.7
10	2	36.7	37.7	44.7	41.4	47.9	43.2
9	3	31.3	54.2	38.9	58.7	42.0	60.8
8	4	24.7	69.1	32.1	73.8	35.0	76.0
7	5	17.5	82.8	24.7	87.5	27.6	89.6
6	6	10.0	95.8	17.2	100.3	20.1	102.2
5	7	2.7	108.7	10.0	112.8	13.0	114.5

		July 21		Aug. 21		Sept. 21	
		Altitude	Azimuth	Altitude	Azimuth	Altitude	Azimuth
12	0	50.7	0.0	42.4	0.0	31.1	0.0
11	1	49.2	21.8	41.2	19.6	29.9	17.4
10	2	45.3	41.7	37.5	38.0	26.7	34.0
9	3	39.5	59.1	32.1	54.6	21.7	49.5
8	4	32.6	74.2	25.5	69.6	15.4	63.9
7	5	25.3	87.9	18.2	83.3	8.4	77.5
6	6	17.8	100.7	10.7	96.3	0.9	90.5
5	7	10.6	113.1	3.4	109.1	− 6.5	103.6

		Oct. 21		Nov. 21		Dec. 21	
		Altitude	Azimuth	Altitude	Azimuth	Altitude	Azimuth
12	0	19.6	0.0	10.3	0.0	6.6	0.0
11	1	18.6	15.6	9.4	14.3	5.7	13.8
10	2	15.7	30.7	6.6	28.3	3.0	27.3
9	3	11.1	45.1	2.3	41.8	− 1.1	40.5
8	4	5.2	58.8	− 3.2	54.7	− 6.6	53.1
7	5	− 1.6	71.9	− 9.8	67.3	****	65.5

Note: To convert from solar time to local time, apply corrections for longitude and equation of time.

APPENDIX C

REQUIRED TEMPERATURE AND ENERGY USE IN SOME AGRICULTURAL AND FOOD PROCESSES

Process	Temperature (°C)	Energy[a]
Sausages and prepared meats		
Scalding, carcass wash, and cleanup	60	46.1
Singeing flame	260	1.12
Edible rendering	93	0.55
Smoking, cooking	68	1.22
Poultry dressing		
Scalding	60	3.33
Natural cheese		
Pasteurization	77	1.35
Starter vat	57	0.02
Make vat	41	0.50
Finish vat	38	0.02
Whey condensing	71—93	10.8
Process cheese blending	74	0.07
Condensed and evaporated milk		
Stabilization	93—100	3.09
Evaporation	71	5.48
Spray drying	177—204	3.78
Sterilization	121	0.57
Fluid milk		
Pasteurization	72—77	1.52
Canned specialties		
Beans		
Precook (blanch)	82—100	0.42
Simmer blend	77—100	0.25
Sauce heating	88	0.21
Processing	121	0.40
Canned fruits and vegetables		
Blanching/peeling	82—100	1.98
Pasteurization	93	0.16
Brine syrup heating	93	1.08
Commercial sterilization	100—121	1.76
Sauce concentration	100	0.46
Dehydrated fruits and vegetables		
Fruit and vegetable drying	74—85	6.16
Potatoes		
Peeling	100	0.35
Precook	71	0.50
Cook	100	0.50
Flake drier	177	1.15
Granule flash drier	288	1.15
Frozen fruits and vegetables		
Citrus juice concentration	88	1.40
Juice pasteurization	93	0.28
Blanching	82—100	2.38
Cooking	77—100	1.49
Wet corn milling		
Steep water evaporator	177	3.86
Starch drier	49[b]	3.20
Germ drier	177	2.08
Fiber drier	538	3.09
Gluten drier	177	1.39

REQUIRED TEMPERATURE AND ENERGY USE IN SOME AGRICULTURAL AND FOOD PROCESSES (continued)

Process	Temperature (°C)	Energy[a]
Steepwater heater	49	0.81
Sugar hydrolysis	132	1.99
Sugar evaporator	121	2.89
Sugar drier	49[a]	0.17
Prepared feeds		
Pellet conditioning	82—88	2.40
Alfalfa drying	204[a]	17.7
Bread and baked goods		
Proofing	38	0.89
Baking	216—238	6.75
Cane sugar refining		
Mingler	52—74	0.62
Melter	85—91	3.48
Defecation	71—85	0.46
Revivification	399—599	4.18
Granulator	43—54	0.46
Evaporator	129	27.84
Beet sugar		
Extraction	60 to 85	4.88
Thin juice heating	85	3.25
Lime calcining	538	3.14
Thin syrup heating	100	7.05
Evaporation	132—138	32.5
Granulator	66—93	0.16
Pulp drier	110—138	17.4
Soybean oil mills		
Bean drying	71	4.27
Toaster desolventizer	102	6.41
Meal drier[a]	177[a]	4.60
Evaporator	107	1.71
Stripper	100	0.32
Animal and marine fats		
Continuous rendering of inedible fat	166—177	17.4
Shortening and cooking oil		
Oil heater	71—82	0.76
Wash water	71—82	0.13
Drier preheat	93—132	0.63
Cooking oil reheat	93	0.34
Hydrogenation preheat	149	0.39
Vacuum deodorizer	149—204	0.37
Malt beverages		
Cooker	100	1.61
Water heater	82	0.56
Mash tub	77	0.63
Grain drier	204[a]	9.68
Brew kettle	100	4.20
Distilled liquor		
Cooking (whiskey)	100	3.33
Cooking (spirits)	160	6.61
Evaporation	121—143	2.45
Drier (grain)	149—204	2.05
Distillation	110—204	8.11

REQUIRED TEMPERATURE AND ENERGY USE IN SOME AGRICULTURAL AND FOOD PROCESSES (continued)

Process	Temperature (°C)	Energy[a]
Soft drinks		
Bulk container washing	77	0.22
Returnable bottle washing	77	1.34
Nonreturnable bottle washing	24—29	0.45
Can warming	24—29	0.55
Cigarettes		
Drying	104[a]	0.45
Rehumidification	104[a]	0.45
Tobacco stemming and redrying		
Drying	104	0.26
Finishing plants, cotton		
Washing	100	16.2
Dyeing	100	4.7
Drying	135	23.4
Finishing plants, synthetic		
Washing	93	37.9
Dyeing	100	16.0
Drying and heat setting	135	24.5
Sawmills and planning mills		
Kiln drying of lumber	149	66.9
Plywood		
Plywood drying	121	53.4
Veneer		
Veneer drying	100	61.0

[a]   No special temperature required. Requirement is simply to evaporate water.
[b]   Process heat used for application, 10^{12} kJ/year.

Values from Peterson, E. A., *ASHRAE Trans. 1979,* Vol. 85, Part I, and *ASHRAE Handbook — 1982 Applications,* ASHRAE, Atlanta, Ga., 1982.

INDEX

S

T

U

V

W

Y

Z